REAL SCIENCE

William C. Kyle, Jr. **Joseph H. Rubinstein** **Carolyn J. Vega**

A Division of The McGraw-Hill Companies

Columbus, Ohio

Authors

William C. Kyle, Jr.
E. Desmond Lee Family
  Professor of Science Education
University of Missouri – St. Louis
St. Louis, Missouri

Joseph H. Rubinstein
Professor of Education
Coker College
Hartsville, South Carolina

Carolyn J. Vega
Classroom Teacher
Nye Elementary
San Diego Unified School District
San Diego, California

PHOTO CREDITS
Cover Photo: © Tim Davis/Tony Stone Images

SRA/McGraw-Hill

A Division of The **McGraw·Hill** *Companies*

Copyright © 2000 by SRA/McGraw-Hill.

Send all inquiries to:
SRA/McGraw-Hill
8787 Orion Place
Columbus, OH 43240-4027

Printed in the United States of America.

ISBN 0-02-683804-4

  5 6 7 8 9 RRW 05

Content Consultants

Gordon J. Aubrecht II
Professor of Physics
The Ohio State University
    at Marion
Marion, Ohio

William I. Ausich
Professor of Geological
    Sciences
The Ohio State University
Columbus, Ohio

**Linda A. Berne, Ed.D.,
    CHES**
Professor/Health Promotion
The University of
    North Carolina
Charlotte, North Carolina

Robert Burnham
Science Writer
Hales Corners, Wisconsin

Dr. Thomas A. Davies
Texas A&M University
College Station, Texas

Nerma Coats Henderson
Science Teacher
Pickerington Local
    School District
Pickerington, Ohio

Dr. Tom Murphree
Naval Postgraduate School
Monterey, California

Harold Pratt
President, Educational
    Consultants, Inc.
Littleton, Colorado

Mary Jane Roscoe
Teacher/Gifted And
    Talented Program
Columbus, Ohio

Mark A. Seals
Assistant Professor
Alma College
Alma, Michigan

Sidney E. White
Professor Emeritus
    of Geology
The Ohio State University
Columbus, Ohio

Ranae M. Wooley
Molecular Biologist
Riverside, California

Reviewers

Stacey M. Benson
Teacher
Clarksville Montgomery
    County Schools
Clarksville, Tennessee

Mary Coppage
Teacher
Garden Grove Elementary
Winter Haven, Florida

Linda Cramer
Teacher
Huber Ridge Elementary
Westerville, Ohio

John Dodson
Teacher
West Clayton
    Elementary School
Clayton, North Carolina

Cathy A. Flannery
Science Department
    Chairperson/Biology
    Instructor
LaSalle-Peru Township
    High School
LaSalle, Illinois

Cynthia Gardner
Exceptional Children's
    Teacher
Balls Creek Elementary
Conover, North Carolina

Laurie Gipson
Teacher
West Clayton
    Elementary School
Clayton, North Carolina

Judythe M. Hazel
Principal and Science
    Specialist
Evans Elementary
Tempe, Arizona

Melissa E. Hogan
Teacher
Milwaukee Spanish
    Immersion School
Milwaukee, Wisconsin

David Kotkosky
Teacher
Fries Avenue School
Los Angeles, California

Sheryl Kurtin
Curriculum Coordinator, K-5
Sarasota County
    School Board
Sarasota, Florida

Michelle Maresh
Teacher
Yucca Valley
    Elementary School
Yucca Valley, California

Sherry V. Reynolds, Ed.D.
Teacher
Stillwater Public
    School System
Stillwater, Oklahoma

Carol J. Skousen
Teacher
Twin Peaks Elementary
Salt Lake City, Utah

M. Kate Thiry
Teacher
Wright Elementary
Dublin, Ohio

UNIT A

Life Science

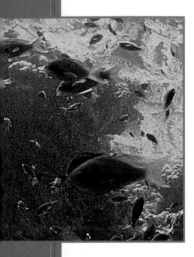

UNIT B

Earth Science

Rocks, soil, and natural resources are
found in Earth's surface layers.

UNIT C

Physical Science

Simple machines help us to do work.

Electricity is a form of energy.

Health Science

Nutrients in food support growth and
good health.

Science Process Skills

Understanding and using scientific process skills is a very important part of learning in science. Successful scientists use these skills in their work. These skills help them with research and discovering new things.

Using these skills will help you to discover more about the world around you. You will have many opportunities to use these skills as you do each activity in the book. As you read, think about how you already use some of these skills every day. Did you have any idea that you were such a scientist?

OBSERVING

Use any of the five senses (seeing, hearing, tasting, smelling, or touching) to learn about objects or events that happen around you.

Looking at objects with the help of a microscope is one way to observe.

COMMUNICATING

Express thoughts, ideas, and information to others. Several methods of communication are used in science—speaking, writing, drawing graphs or charts, making models or diagrams, using numbers, and even body language.

Making a graph to show the rate of growth of a plant over time is communicating.

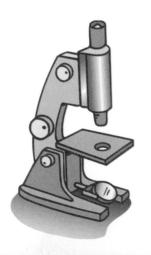

CLASSIFYING

Organize or sort objects, events, and things that happen around you into categories or groups. The classified objects should all be alike in some way.

Sorting students in the room into groups according to hair color is classifying.

Using Numbers

Use math skills to help understand and study the world around you. These skills include ordering, counting, adding, subtracting, multiplying, and dividing.

Comparing the temperatures of different locations around your home is using numbers.

Measuring

Use standard measures of time, distance, length, area, mass, volume, and temperature to compare objects or events. Measuring also includes estimating and using standard measurement tools to find reasonable answers.

Using a meterstick to find out how far you can jump is measuring.

Constructing Models

Draw pictures or build models to help tell about thoughts or ideas or to show how things happen.

Drawing the various undersea formations on the ocean floor is constructing a model.

Inferring

Use observations and what you already know to reach a conclusion about why something happened. Inferring is an attempt to explain a set of observations. Inferring is not the same as guessing because you must observe something before you can make an inference.

Imagine you put a lettuce leaf in your pet turtle's aquarium. If the lettuce is gone the next day, then you can **infer** that the turtle ate the lettuce.

PREDICTING

Use earlier observations and inferences to forecast the outcome of an event or experiment. A prediction is something that you expect to happen in the future.

Stating how long it will take for an ice cube to melt if it is placed in sunlight is **predicting.**

INTERPRETING DATA

Identify patterns or explain the meaning of information that has been collected from observations or experiments. Interpreting data is an important step in drawing conclusions.

You interpret data when you **study** daily weather tables and **conclude** that cities along the coast receive more rainfall than cities in the desert.

IDENTIFYING AND CONTROLLING VARIABLES

Identify anything that may change the results of an experiment. Change one variable to see how it affects what you are studying. Controlling variables is an important skill in designing investigations.

You can **control** the amount of light plant leaves receive. Covering some of the leaves on a plant with foil allows you to compare how plant leaves react to light.

HYPOTHESIZING

Make a statement that gives a possible explanation of how or why something happens. A hypothesis helps a scientist design an investigation. A hypothesis also helps a scientist identify what data to collect.

Saying that bean seeds germinate faster in warm areas than cold areas is a hypothesis. You can **test** this hypothesis by germinating bean seeds at room temperature and in the refrigerator.

DEFINING OPERATIONALLY

An operational definition tells what is observed and how it functions.

Saying the skull is a bone that surrounds the brain and is connected to the backbone is an operational definition.

DESIGNING INVESTIGATIONS

Plan investigations to gather data that will support or not support a hypothesis. The design of the investigation determines which variable will be changed, how it will be changed, and the conditions under which the investigation will be carried out.

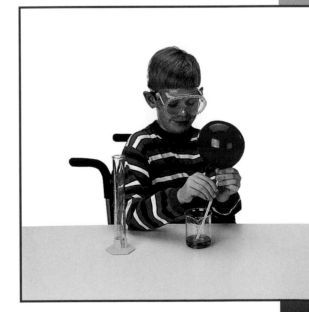

You can **design an investigation** to determine how sunlight affects plants. Place one plant in the sunlight and an identical plant in a closet. This will allow you to control the variable of sunlight.

EXPERIMENTING

Carry out the investigation you designed to get information about relationships between objects, events, and things around you.

Experimenting pulls together all of the other process skills.

UNIT A

Life Science

1

Organisms and Where They Live

Life on water suits many living things. Beavers build dams on rivers and lakes. Alligators dig "gator" holes with their five webbed toes. Water lilies bloom and grow floating on ponds.

Other living things survive in different kinds of places. Plants and animals in an Arizona desert live in the extreme heat with little water. On some mountains, plants and animals live in cool temperatures. In an ocean, plants and animals must be able to survive under water. In all of these places, nonliving things exist too. Rocks, soil, and air are nonliving things that are useful for living things.

The Big IDEA

Organisms are adapted to their environments.

CHAPTER SCIENCE INVESTIGATION

Make a model of an ecosystem to see how plants and animals get what they need to live and grow. Find out how in your *Activity Journal.*

Habitats of Organisms

Find Out

- What an environment is
- What a habitat is
- How organisms meet their needs
- What makes up an ecosystem
- What communities and populations are

Vocabulary

organisms
environment
habitat
ecosystem
community
population

The Big QUESTION

How do animals and plants live where they do?

Can you imagine polar bears without fur? Fish without fins? Spiders without legs? Birds without wings? Living things have body parts and features that help them survive. They get everything they need from the places they live. In this lesson, you will learn how living things survive. You will learn how plants and animals depend on other living and nonliving things.

Living Things and Their Environments

Organisms (or′ gə niz′ əms) are living things. All plants and animals are organisms. An **environment** is made up of all of the living and nonliving things that surround an organism. To survive, every living thing uses things from its environment.

A4

For example, the roots of plants use minerals and water from the soil. Beavers build dams with logs and branches from trees. Birds use twigs and grass to build nests.

Humans affect their environment too. Simply walking on the grass changes or affects that living thing—the grass. If you plant a garden, you change that environment. Humans, like all living things, affect or interact with their environments to meet their needs.

Habitats

Like you, plants and animals have needs. They need air, water, shelter, food, and enough space to move and grow. A place where living things live and grow is called a **habitat.** The habitat meets the needs of the living things within it.

Animals often use plants for shelter as well as food. Elf owls such as the one shown here live inside saguaro cacti.

Desert animals such as this tortoise are adapted to dry environments. Plants can provide food and water to desert animals.

Many different kinds of organisms can share the same habitat. This backyard is filled with living things.

Organisms Meeting Their Needs

Organisms are adapted to the places they live. A lizard needs a warm habitat. It could not live near the north pole. A polar bear is adapted to a cold climate. Thick fur and a layer of fat beneath the fur keep polar bears warm. A polar bear could not live in the lizard's warm habitat.

Every habitat provides the food that the living things need. In their natural habitat, polar bears eat seals and fish. Modern zoos do their best to copy animals' natural habitats. Polar bears in zoos are fed the kind of food they would find in their natural habitats.

Living things also use nonliving things from their habitats. For example, rocks, ice, and snow are nonliving things that are part of many animals' habitats. Bears often live in large areas of rock called dens when it is very cold outside. In the winter, deer may eat ice and snow to get the water they need. In hot desert environments, lizards and snakes often crawl under rocks when they need shelter from the heat of the sun.

Polar bears

Both living and nonliving things make up an ecosystem.

Ecosystems

A system is a group of things that work together. An **ecosystem** (ēk′ ō sis′ təm) is made up of groups of living things and their habitats. Ecosystems include all of the living and nonliving things within a habitat.

An ecosystem can be large or small. It can be as large as an ocean. It can be as small as a decaying log. A log is often home to ants, termites, and other insects. They depend on the log for shelter. As the insects burrow holes in the log, the log falls apart. As it falls apart, the log becomes part of the soil. Plants can grow in the soil. The plants provide food to animals in the area. All these things in the ecosystem depend on one another in some way.

Communities and Populations

The plants and animals that live in the log are part of a community. A **community** is all of the living things in an area. A group of the same kind of living things in the same area is called a **population** (pop′ yə lā′ shən). The ants make up one population. The other kinds of insects make up other populations. The plants and animals each have their own populations. Each community has different populations.

Several populations often share the same habitat. A tree can be a habitat for birds, squirrels, insects, and even other plants. A pond or lake can be home to hundreds of different kinds of plants and animals.

Populations that share the same ecosystem often have the same kinds of needs. Fish, crabs, and sea urchins can live in the same ocean habitat. They all get the food they need from the tiny plants and animals in the water.

People and animals share habitats.

Animals' body parts are adapted to help the animals get what they need from ecosystems. A fish uses its fins to steer and move through the water. Crabs use their claws to catch and hold food. Cows and horses use their strong teeth to tear off and chew the grass they eat.

Plants also have parts that are adapted to their different habitats. Many rain forest plants have large, wide leaves that collect sunlight in shady environments. Desert plants often have short, thick stems or leaves that store water to use during dry periods. All organisms have specialized parts that allow them to live in the ecosystems around them.

A community of gazelles, zebras, giraffes, and gemsboks sharing a habitat in Namibia

CHECKPOINT

1. What is an environment?
2. What is a habitat?
3. Name three basic needs that organisms have.
4. What makes up an ecosystem?
5. What is a community? A population?
 How do animals and plants live where they do?

ACTIVITY

Observing Part of an Ecosystem

Find Out

Do this activity to see what kinds of living and nonliving things make up a small ecosystem.

Process Skills

Observing
Classifying
Communicating

WHAT YOU NEED

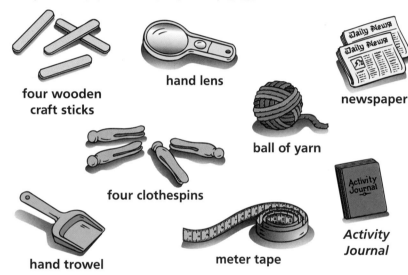

four wooden craft sticks

hand lens

newspaper

four clothespins

ball of yarn

hand trowel

meter tape

Activity Journal

WHAT TO DO

1. Choose a natural area to study. Use the clothespins and yarn to mark off a square that is 1 m on each side.

   Safety! *Check the area for animals or plants that might be harmful.*

2. Try not to disturb the area you are observing. Use craft sticks to look between the plants. Look for signs of animals, such as holes or footprints.

3. Use the hand trowel to dig up some of the topsoil. Spread it on the newspaper. Use the hand lens to **observe** the topsoil.

4. Draw a picture of some of the things you find in the soil. **Classify** them as living or nonliving. **Record** the number and kind of each organism that you see.

CONCLUSIONS

1. How many different kinds of nonliving things did you have in your ecosystem? What nonliving things did you have the most of?

2. How many different kinds of living things did you have in your ecosystem? What living things did you have the most of?

3. Is anything in your ecosystem eating something else? How can you tell?

ASKING NEW QUESTIONS

1. What other things might live in the ecosystem you observed?

2. Could you live in this ecosystem? Could a polar bear? Why?

SCIENTIFIC METHODS SELF CHECK

✔ Did I **observe** the topsoil with a hand lens?

✔ Did I **classify** the living and nonliving things in the ecosystem by drawing pictures of them?

✔ Did I **communicate** my findings by writing them down?

Plants and Animals Depend on Each Other

Find Out

- How some animals catch other animals to eat
- What food chains and food webs are
- How producers differ from consumers

Vocabulary

predators
prey
food chain
food web
producers
consumers
scavengers
decomposers

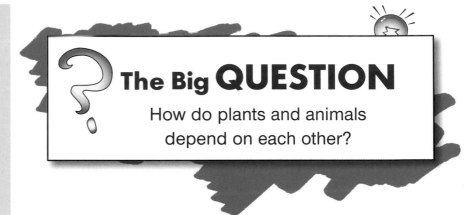

The Big QUESTION

How do plants and animals depend on each other?

Nothing in nature is wasted. Every living and nonliving thing has a role. Populations in ecosystems interact and compete to get the food they need to survive. This lesson introduces you to the relationships of living things within ecosystems. You will also learn about food chains.

Predators and Prey

Except for animals living in people's homes and in zoos, animals must find their own food. Those that eat meat must eat other animals. Animals that eat other animals are called **predators** (pred′ ə tərs). The animals they capture and eat are called **prey** (prā). The predator stalks, or "preys upon," the animal it captures and eats.

Lions prey on zebras. Hawks prey on mice. Sharks prey on smaller fish. Almost all living things are preyed upon by other living things. All animals live in places where they can find what they need. Predators and prey share the same habitats.

Predators are doing what is natural. Even a well-fed pet cat will eat mice and birds. The mice and birds, in turn, eat other living things.

What predator/prey relationships do you see in this picture?

Food Chains

All living things are part of a food chain. A **food chain** is the way food energy passes from one organism to another in a community. The different food chains in a community link together to form a **food web.** Predators often eat more than one kind of prey.

Nature does not "give" meals to every animal. They must compete for limited resources such as food, water, and shelter. Different kinds of animals compete with one another. So do animals from the same population. Each competes to get what it needs to live.

Competition for food and water can change the size of a population. If there is not enough food, the weaker members of a population don't get enough to eat. They may not survive because the stronger animals eat all the food.

The arrows show how the energy is moving.

Competition for food is one way nature limits the size of a population. Disease is another way. Diseases kill the weaker animals. The stronger animals survive.

Like you, all living things get energy from food. Living things need energy to survive. It also allows them to move, grow, and reproduce.

Follow the arrows to see the beginning and end of each food chain. How many food chains are there?

What else might the hawk eat from another food chain?

Overlapping food chains are called a *food web*. What else might the fox eat from another food chain?

Producers and Consumers

Energy moves through food chains and food webs. All energy on Earth comes from the sun. Plants use the sun's energy to produce—or make—food energy. Within the food chain, plants are called **producers.**

Animals that eat plants consume—or use—the food energy stored in plants. Animals that eat meat also consume plant energy. How? By eating the animals that *do* eat plants. In a food chain, animals are called **consumers.**

Here's how it works. A plant-eating animal eats a green plant. The plant contains food energy made from the sun's energy. The plant is the first link in the food chain. The animal that eats the plant is the second link. Some of the sun's energy is passed on to the animal in the food.

When a meat-eating animal eats the plant-eating animal, energy is again passed along. The meat-eater is the third link in the food chain.

Food chains and webs all depend on the sun. If the sun were any closer or any farther away from Earth, producers might grow faster, or slower, or not at all.

A change in the number of producers would affect all the food chains and webs—every living thing on Earth, including you!

Pathway of Energy

Some energy passes from producers to consumers. But as energy continues to move, a large part of it is used by living things.

Energy used

Energy passed on

Energy used

Energy passed on

Energy

Some consumers are **scavengers** (skav′ ən jərs). Vultures and other scavengers survive by eating dead organisms. Worms and some other organisms are decomposers. **Decomposers** (dē′ kəm pō′ zərs) get energy by causing dead organisms to decay.

Just as animals depend on plants for food energy, many plants depend on animals for their survival too. In order to survive and produce offspring, many kinds of plants need to spread their seeds and pollen. Because plants cannot move from place to place on their own, seeds and pollen must be carried by wind, rain, or animals. Many different kinds of animals help spread seeds and pollen. Bees spread pollen as they fly from flower to flower. Some seeds get stuck in an animal's fur and may travel with the animal for a long time before falling off and growing in a different place.

Every organism in an ecosystem helps other living things to survive.

These vultures are scavengers.

CHECKPOINT

1. How are predators and prey different?
2. What is a food chain?
3. What is a producer? A consumer?
 How do plants and animals depend on each other?

ACTIVITY
Making a Food Chain

Find Out

Do this activity to see how energy is transferred from producers to consumers.

Process Skills

Classifying
Communicating
Constructing Models
Inferring

WHAT YOU NEED

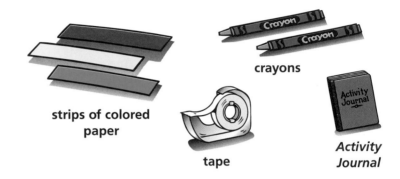

strips of colored paper

crayons

tape

Activity Journal

WHAT TO DO

1. **Write** the names of one producer (plant) and two or three consumers (animals) that make up a food chain.

2. **Write** each name on a separate strip of paper.

3. Place the strips of paper in front of you. **Sort** them to show the order of energy flow in a food chain.

4. Carefully tape the ends of the first plant or animal in your food chain so that it forms a circle. Make sure the name shows on the outer part of the loop.

5. Loop the next strip of paper in the food chain through the loop and tape its ends together to make a second loop. Do this again with the other strips to create all of the links in your food chain. Be careful to add the links in the same order as your sorted paper strips.

6. **Explain** your food chain model to a classmate. Why did you choose the plants and animals that you did? **Record** your answers.

CONCLUSIONS

1. Why did you have to use the names of both plants and animals for your food chain model?

2. What different names did your classmates choose for their models?

3. Would your food chain model still be a "chain" if you had used plant names only?

ASKING NEW QUESTIONS

1. What would happen to your model if you removed one of the middle links from the chain?

2. In a real food chain, what happens to the consumers if all the producers disappear?

SCIENTIFIC METHODS SELF CHECK

✔ Did I **classify** the plants and animals that I chose as either producers or consumers?

✔ Did I **model** a food chain with the paper strips?

✔ Did I **explain** my model to a classmate?

Organisms Adapt

Find Out

- What an adaptation is
- What adaptations organisms have for meeting their needs
- What inherited and learned characteristics are

Vocabulary

behavior
adaptations
camouflage
reproduce
inherited
  characteristics
learned characteristics

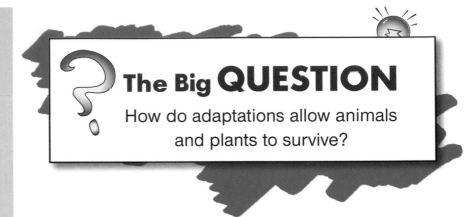

The Big QUESTION
How do adaptations allow animals and plants to survive?

What if you had feet like a bird? Or instead of hands you had fins like a fish? It would be hard for you to walk—and impossible for you to hold a fork to eat your food.

You have the kind of hands you need. You have the kind of feet you need. All your characteristics are adapted to your environment. Birds, fish, plants, and all other organisms have characteristics adapted to their environments.

Adaptations

Living things are adapted to their environments. An animal's body parts and the parts of a plant are called *structures*. How a living thing acts in its environment is called **behavior. Adaptations** (ad′ əp tā′ shəns) are the ways structures or behaviors help living things stay alive.

Adaptations help living things grow, have young, and survive in their environment.

All plants and animals get their food, water, and shelter from Earth. They respond to changes in Earth's weather, soil, and water. Over time, each kind of organism has adapted to the place on Earth where it lives. Living things that do not adapt to their environment will not survive.

Auk—A diving bird that lives on northern seacoasts. It has webbed feet and short wings that it uses as paddles for swimming.

Angelfish—A colorful fish with fins shaped like wings. Its fins help it glide through the water. It lives in warm, tropical waters.

Adaptations for Meeting Needs

Aardvark

The saguaro cactus is adapted to life in the desert.

Different organisms have different adaptations. The adaptations let the organisms meet their needs within their habitats. The aardvark's natural home is in Africa. Aardvarks have long tongues that they use to catch and eat termites and ants. Aardvarks could not live in the Arctic. The weather, food, and climate are too different. The aardvark is adapted to life in Africa.

Desert plants are adapted to the warm, dry climates of hot deserts. For example, the saguaro cactus can store water in its thick stems to use during long periods when there is no rain. Because the saguaro is adapted to live in the desert, it could not survive in a cold, wet place like a marsh.

Some animals have adaptations that help to protect them from being caught and eaten by other animals. The prairie dog lives in open grasslands where it may be hunted by large birds, coyotes, snakes, or other animals. To stay hidden from these predators, the prairie dog has an adaptation called **camouflage**—coloring, shape, and size that help some animals blend in with their environment. Because the prairie dog's

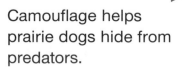

Camouflage helps prairie dogs hide from predators.

fur is the same color as the tall grasses growing in its habitat, the prairie dog is hard to see when it is not moving.

Scavengers are organisms that are adapted to live on "leftovers." Vultures, crows, hyenas, and many other animals eat what is left after other creatures have eaten what they want. Scavengers such as worms, starfish, and other small organisms live in soil or on the bottom of oceans, lakes, and ponds. They eat the dead organisms in the soil or those that drop from the waters above.

All organisms **reproduce,** or have young of their kind. Some plants reproduce by scattering seeds. Others reproduce by forming bulbs or underground stems. Some animals give birth. Others lay eggs. Humans usually give birth to one baby at a time. A fish may lay thousands of eggs. Each organism is adapted to reproduce itself in a way that gives its young a good chance of survival.

The crow is a scavenger.

This starfish is adapted to its ocean habitat.

Inherited and Learned Characteristics

The abalone has adapted to life in an underwater habitat.

Organisms have different ways to stay alive. Some move fast. They outrun predators. Some organisms have camouflage to help them hide from predators. Some organisms taste bad. Predators stay away from them. Others have bristles or hard shells to protect their bodies. These adaptations are all examples of inherited characteristics. **Inherited characteristics** are adaptations that living things inherit from their parents.

Learned characteristics are characteristics that are taught. For example, mother bears teach their cubs how to find food. Birds teach their young to fly and to catch prey. Prairie dogs teach their pups to call or "bark" warnings to others when a predator is nearby.

Bears teach their young to find food.

Maple tree in autumn

Both learned and inherited characteristics help organisms stay alive. These characteristics allow organisms to respond to their environments. A squirrel responds to fall weather by burying and storing nuts. This response helps it survive by providing food during the winter. A maple tree responds to fall weather by shedding its leaves. It responds to long spring days by growing leaves again, which make the food it needs. Its response, too, helps it stay alive.

CHECKPOINT

1. What is an adaptation?
2. What are some adaptations organisms have for meeting needs?
3. What is an inherited characteristic? A learned characteristic?

[?] How do adaptations allow animals and plants to survive?

ACTIVITY
Watching Worms

Find Out

Do this activity to see how earthworms are adapted to live in soil.

Process Skills

Observing
Communicating

WHAT YOU NEED

earthworm

waxed paper

soil

shallow box

hand lens

Activity Journal

WHAT TO DO

1. Line the box with waxed paper.
2. Gently place the worm on the paper.
3. **Observe** the worm's movements for several minutes. **Record** descriptions of the movements.

4. Use the hand lens to look at the worm's body sections. Find the wiry hairs on the body sections.

5. Sprinkle a layer of soil on top of the waxed paper. Place the worm on the soil. Again, **observe** the worm's movements for several minutes and **record** them.

6. Repeat the activity, again recording your observations.

CONCLUSIONS

1. Over which surface does the worm move better?

2. What adaptation helps an earthworm move over surfaces?

3. Were your observations the same the second time you made them?

ASKING NEW QUESTIONS

1. Do you think the earthworm behaved the same way in the box with soil as it would in its natural habitat?

2. How are an earthworm's wiry hairs like the fins of a fish?

SCIENTIFIC METHODS SELF CHECK

✔ Did I **observe** the worm's movements on both surfaces?

✔ Did I **observe** the worm's body through the hand lens?

✔ Did I **record** my observations?

Review

Reviewing Vocabulary and Concepts

Write the letter of the word or phrase that completes each sentence.

1. Another name for a living thing is ___.
   a. an ecosystem   **b.** a community
   c. an organism   **d.** a food web

2. A place where living things live and grow is called ___.
   a. a habitat   **b.** a population
   c. camouflage   **d.** a food chain

3. Predators, scavengers, and decomposers are energy ___.
   a. consumers   **b.** communities
   c. chains   **d.** producers

4. A change in structure or behavior that helps a living thing stay alive is called ___.
   a. a behavior   **b.** an adaptation
   c. inherited characteristics   **d.** a learned characteristic

Match the definition on the left with the correct term.

5. all of the animals and plants that live in an area   **a.** prey

6. animals that are captured and eaten by other animals   **b.** learned characteristics

7. living things that use the sun's energy to produce food energy   **c.** community

8. traits or behaviors that are taught   **d.** producers

Understanding What You Learned

Write the letter of the word or phrase that completes each sentence.

1. Energy from the sun is passed from plants to animals through ___.
   - **a.** diseases
   - **b.** learned characteristics
   - **c.** zoos
   - **d.** the food chain

2. Comparing a predator and its prey is similar to comparing a scavenger and ___.
   - **a.** dead animals
   - **b.** decomposers
   - **c.** plants
   - **d.** learned characteristics

3. Animals get energy from ___.
   - **a.** cooperation
   - **b.** food
   - **c.** camouflage
   - **d.** water

4. Living things that cause dead plants and animals to decay are called ___.
   - **a.** producers
   - **b.** scavengers
   - **c.** decomposers
   - **d.** prey

Applying What You Learned

1. How is a polar bear adapted to its cold habitat?

2. Describe how disease and competition for food and water affect the size of a population.

 3. What are some ways that organisms adapt?

For Your **Portfolio**

Think about how animals and plants are adapted to survive within their environments. Now write a short description of how each of these human characteristics helps us in our daily lives: thumbs, eyelids, hair, and tongue.

Humans in the Ecosystem

Earth is one big ecosystem. It seems to have enough water, oxygen, energy, and space for every living thing. But will Earth always have enough of these things?

People all over the world have growing needs for food and places to live. In many areas, land, water, and building supplies are hard to find. People may use so much of certain materials that other living things do not have enough of what they need to live. As you will discover, people make choices every day. Some of these choices may be good or bad for other living things.

The Big IDEA

Humans affect the environment all over the world.

CHAPTER **SCIENCE** **INVESTIGATION**

Learn how decomposition of a plant part might affect an ecosystem. Find out how in your *Activity Journal.*

People's Place in the Ecosystem

Find Out

- How people interact with other living things
- What people's basic needs are
- How people get the things they need

Vocabulary

carbon dioxide
natural resources
recycle

The Big QUESTION

How do people fit into the ecosystem?

*H*ave you ever thought about living in a place that is different from where you live now? If you lived in Tanzania, you could explore the rain forest. In Austria, you might climb the Alps, and in Nepal, you might climb Mount Everest. In southern South America, you even might take a trip to nearby Antarctica. But no matter where you live, one thing is true. You share your environment with many different kinds of living things.

People Affect Their Environment

People all over the world live with other organisms in their environment. In doing so, they help meet each other's needs. Trees and other plants give off the oxygen people need to breathe. By planting trees, people can improve their environment. The things that people do can also hurt other organisms and the ecosystems around them. If a factory is built next to a river, the factory may hurt the environment. Wastewater from the factory may get into the river. The wastewater may cause the fish in the river to become sick. People who catch and eat the fish may also get sick. People and the fish are connected. Because they interact within their environment, they have a relationship with each other.

Tokyo, Japan
Like all living things, people in this crowded city depend on Earth's ecosystem for the things they need to live.

In this fish tank, the plant and the fish have a relationship. The plant gives off oxygen, a gas the fish needs to live. The fish gives off **carbon dioxide** (kär´ bən dī ok´ sīd), a gas the plant needs. If the right amounts of carbon dioxide and oxygen are in the water, the needs of both the plant and the fish are met.

A33

Using Natural Resources

A wheat field

A vegetable garden

Think about some of the things you use every day. Is your shirt made of cotton? Cotton comes from a plant. Is your chair or desk made of wood? How many of the things in your home are made of wood? Wood comes from trees. Cotton and trees are examples of natural resources. **Natural resources** are things found in nature that are useful to and needed by living things.

Trees, cotton plants, and all other living things need water. Water is an important natural resource. Besides water for drinking, we need water for cooking, bathing, and brushing our teeth. Some factories need water to run machines and make power. Fortunately, water is a resource that we can **recycle**—we can use it over and over. Rain brings freshwater to rivers, lakes, and other places where humans get their water. In some places, there are factories that recycle water for human use. Some factories clean river water so that it is safe for people to use in their homes. Others make freshwater from seawater by taking the salt out of it.

Soil as a Resource

Today, throughout the world, more and more people are living in cities. There is one important resource many of these people may not think about often. It is the resource in which most of our food grows. That resource is soil.

Soil covers much of the top layer of Earth. Soil is made up of tiny pieces of rock, some minerals like calcium, and humus— bits of decayed animals and plants. Soil contains nutrients and holds water for plants. Depending on what type it is, soil may hold a lot of water or very little water. Sandy soil can be found in many places, not just in deserts. Plants that grow in sandy soil have roots that allow the plant to live in a dry environment. Clay is a heavy, thick soil that contains many minerals. Some kinds of plants grow best in clay. Black or brown "topsoil" is rich in many nutrients, making it suitable for growing many of the plants that humans use for food.

Sandy soil, clay, and topsoil have one thing in common as natural resources. They can be washed away by rain or flooding, ending up in rivers and oceans. Once there, the soil is lost forever. The roots of trees and other plants can hold soil in place so that it won't wash away too quickly. If all of the trees and other plants in a certain area are cut down, that area's soil will be washed away by the rain and blown away by the wind. Soil is a resource that people depend on. It is important to preserve because it cannot be recycled.

A tree farm

The Air We Breathe

Take a deep breath. The air that fills your lungs when you breathe contains oxygen. Do you remember the plant in the fish tank? It gave off the oxygen that the fish needed to live. Green plants do the same for people. They supply the air with the oxygen we need. When we breathe out, we give off carbon dioxide, a gas that green plants need. Because people and plants both need a mixture of oxygen and carbon dioxide in the air, people and green plants depend on each other.

When you think of Earth as an ecosystem, you can understand why the rain forests are important. The rain forests are huge areas of trees and other plants that give off oxygen. In some areas of the rain forest, people have cleared the land of trees so they can grow crops or raise livestock. Because they are an important part of our planet's ecosystem, the clear-cutting of the rain forests affects living things all over Earth.

Earth is rapidly losing rain forests.

Earth's Changing Habitats

Years ago, most people lived on farms. They depended on the land to get what they needed. Today things are different. Most people live in cities. They depend on machines and businesses to get some of the things they need, but they still need natural resources to survive. There are also many more people on Earth today than ever before. As the number grows, more and more people use natural resources.

Earth is changing every day. Some of the changes are natural. Some of the changes are caused by the actions of people. As people use resources, they can affect Earth in harmful ways. They may use too much of a resource such as water. Or they may pollute the air, land, and water.

People are learning more about the ways they affect their environment. Scientists are finding ways that we can protect Earth. As you learn about the environment, you will discover some smart things you can do to help Earth.

The trees these children are planting will help Earth for years to come.

CHECKPOINT

1. How do people interact with other living things?

2. What are people's basic needs?

3. How do people get the things they need?

[?] How do people fit into the ecosystem?

ACTIVITY

Sampling Soil

Find Out

Do this activity to see the differences between two kinds of soil.

Process Skills

Observing

Communicating

WHAT YOU NEED

two soil samples in two paper cups

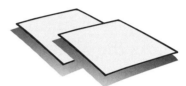

two sheets of manila drawing paper

hand lens

plastic spoon and knife

Activity Journal

WHAT TO DO

1. Fold a piece of paper in half and lay it flat. Place a spoonful of soil from one cup on one side of the paper. Place a spoonful from the other cup on the other side of the paper. Label one sample "A" and the other sample "B." What do you think you will find out about the soil?

Safety! *Be sure to keep your fingers out of your mouth while you do this activity. When you finish, wash your hands.*

A38

2. Use the knife to spread out the soil. **Observe** both samples with your hand lens. Pull to the side any small pieces of rock, plant, or animal matter you see. Pick up some soil with your fingers and feel it. Compare the two samples and observe their differences.

3. **Record** what you observe. Include:
   • color (red, yellow, brown, black)
   • texture (how it felt—sticky, gritty, wet, cool)
   • contents (any small pieces of rock, plants, or animal matter you pulled to the side)

4. Would you expect the same results if you did this activity in another part of the country?

CONCLUSIONS

1. How were the soil samples different?
2. What is soil made of?
3. Did you see any plant or animal matter in either of your soil samples?

ASKING NEW QUESTIONS

1. What else can you learn about soil on the Internet or in the library?
2. Check out one of the sources you found in Question 1. List three new things you learned about soil from your source.

SCIENTIFIC METHODS SELF CHECK

✔ Did I **observe** differences in the samples?

✔ Did I clearly **record** what I saw?

Taking Care of Earth

Find Out

- Why conservation is important
- What happens to the things we throw away
- How humans can work to save natural resources

Vocabulary

conservation
environmentalists
biodegradable
compost

The Big QUESTION

What can you do to protect Earth's resources?

As you know, living things find the air, water, food, and space they need to live on Earth. They use Earth's resources. Many of these natural resources, such as coal and oil, often cannot be replaced or reused. It takes millions of years for them to form. Other resources, like aluminum and copper, can be recycled— they can be used more than once.

People Who Care for Earth

What people do to Earth's land, water, and air may cause problems for other living things. When people practice **conservation** (kon′ sər vā′ shən), they work to protect Earth and its resources. These people are called conservationists or **environmentalists** (en vī′ rən men′ tə lists).

We can all become environmentalists by working to protect the environment. How can we protect the resources we have?

Many people are working on new ways to save Earth's resources. Some try to find ways to clean up pollution in our water supply. Others help people learn to use less water in their homes and businesses.

People are also working on reducing pollution in the air. Air pollution from cars and factories has been a problem for a long time. Now we know that saving Earth's trees and forests is important too. They give off the oxygen we need to breathe.

There are simple ways for you to become an environmentalist. You can save water by turning off the faucet when you brush your teeth. You can also help save trees by using both sides of a piece of paper before throwing it away. Because paper is made from trees, reusing and recycling paper helps conserve them. Throughout your day, think about things you might do to save natural resources.

These people are working together in their own neighborhood to take care of Earth.

Environmentalists are interested in how the smoke coming from these chimneys affects the ecosystem.

Garbage, Garbage Everywhere!

Do you ever wonder what happens to the garbage you throw away? It seems to just disappear. But some of your garbage will be around for a very long time. Most garbage is collected each week and taken to a landfill. Here it is placed in the ground and covered with dirt. This may seem all right, but what happens when we throw out too much garbage? In many places there is not enough land to bury it all.

Some garbage will not harm the environment. Materials that were made from animals or plants are **biodegradable** (bī′ ō di grā′ də bəl). Biodegradable garbage will decay—break down into smaller and smaller parts—and become part of Earth. For example, the apple cores, bread crusts, and eggshells you throw away will decay naturally. If you bury them in the ground or pile them up in a garden, these biodegradable materials will become part of the soil. This rich soil mixture is called **compost.** Compost can be mixed in with the soil in gardens to help new plants grow.

As organisms decay, heat is produced. This heat makes biodegradable materials break down and become compost very quickly. Most garbage, such as plastic bags, soft-drink cans, and candy wrappers, is not naturally degradable. These things will not become compost. They will take up space in landfills for many, many years. However, there are ways that people can reduce the amount of garbage that goes into landfills.

This landfill is almost full.

These volunteers are picking up trash in a park close to home.

Cleaning Up Earth

One way to reduce the amount of garbage is by recycling. You know that recycling is reusing material or changing it into something that can be used again.

Most paper, plastic, glass, and aluminum can be recycled. You can find out what materials your community recycles. If your school does not have a recycling program, maybe you can help start one.

People all over the world are helping to clean litter from beaches, roadsides, and parks. By picking up litter, you can help keep your environment clean and safe.

People are also learning better ways to get rid of harmful chemicals. Paint, antifreeze, and cleaning materials are made of chemicals. Some of the chemicals can hurt living things. If the chemicals are dumped into the ground, they can get into the water supply. Chemicals like these need to be thrown out in a safe way that keeps them from harming the environment.

Taking Care of Your Environment

You have learned to think of Earth as one big ecosystem. All living and nonliving things are a part of it. When people hurt the ecosystem, they hurt other living things. Taking care of Earth is a big job. None of us can do it alone. Here are some other things you can do to help.

In Your Home

Save glass jars and bottles, newspapers, and aluminum cans. Help take them to a recycling center.

Reuse plastic bags to line wastebaskets.

Use rags instead of paper towels.

Take a short shower instead of a bath to save water.

To save energy, turn off any lights you don't need.

At School

Bring your lunch in a reusable container instead of a paper bag.

If you use computers in your classroom, try to reuse the computer paper for other projects.

Try to use recycled paper whenever possible.

People can help recycle Earth's resources.

Other Ways You Can Help

Take your own shopping bags to the store. Reuse the shopping bags until they wear out.

Read labels. Look for the word *biodegradable.* Try to use biodegradable cleaning materials. These will not permanently poison the soil or water.

Work with an adult to pick up trash and litter in your neighborhood. Every little bit makes a big difference.

Talk to your family, friends, and neighbors about ways they can help save Earth's resources. Share what you have learned!

These are just a few of the ways you can protect your environment. If you want to learn more about how to protect Earth, write to some environmental protection groups and ask for information. You can find the addresses at your local library.

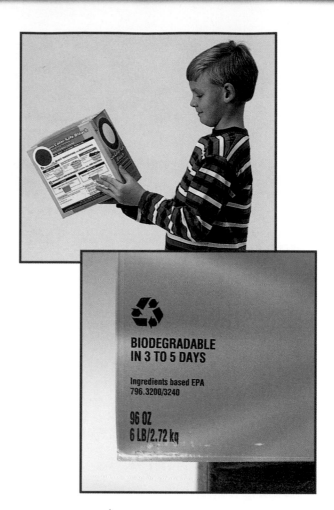

A label will say whether soap has biodegradable chemicals in it.

CHECKPOINT

1. Why is conservation important?
2. What happens to the things we throw away?
3. What are humans doing to save resources?

 What can you do to protect Earth's resources?

ACTIVITY

Recycling Paper

Find Out

Do this activity to learn how to conserve a natural resource by recycling paper.

Process Skills

Observing
Classifying
Predicting
Using numbers

WHAT YOU NEED

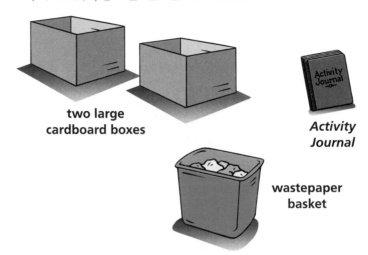

two large
cardboard boxes

Activity Journal

wastepaper basket

WHAT TO DO

1. Work with a small group of other students. Put all of your group's wastepaper in the wastepaper basket every day for a week.

2. Label one box "Reuse." Label the other box "All Used Up."

3. At the end of the day, **observe** and **classify** the paper from the wastepaper basket. Put all the paper that has only been used on one side in the "Reuse" box. Put the rest of the paper in the "All Used Up" box.

Safety! *Be careful when handling the paper. Paper can cut your fingers.*

4. **Predict** which box will have more paper at the end of the week.

5. At the end of the week, **count** and see how many sheets of paper are in each box.

6. **Divide** the paper in the "Reuse" box among members of your group. Use that paper for school assignments.

7. Take the paper from the "All Used Up" box to a place where it can be recycled.

CONCLUSIONS

1. At the end of the week, see if your prediction was correct.

2. What natural resource have you conserved?

3. How much paper might be saved in one school year?

ASKING NEW QUESTIONS

1. Are there other things you can do in your classroom to conserve resources like trees?

2. How might you find out about the recycling habits of schoolmates or neighbors?

SCIENTIFIC METHODS SELF CHECK

✔ Did I make a **prediction** about the amount of paper in each box?

✔ Did I **confirm** that one box had more paper in it?

✔ Did I **classify, divide,** and **count** the paper in each box?

Review

Reviewing Vocabulary and Concepts

Write the letter of the word or phrase that completes each sentence.

1. Materials found in nature that are useful or needed by living things are ___.
   - **a.** trees
   - **b.** relationships
   - **c.** natural resources
   - **d.** environmentalists

2. A gas that animals need is called ___.
   - **a.** a chemical
   - **b.** conservation
   - **c.** oxygen
   - **d.** compost

3. A person who works to protect the environment is ___.
   - **a.** an environmentalist
   - **b.** compost
   - **c.** a resource
   - **d.** a chemical

4. The natural resource from which people get wood is ___.
   - **a.** trees
   - **b.** relationships
   - **c.** water
   - **d.** soil

Match the definition on the left with the correct term.

5. working to protect Earth's resources
   - **a.** biodegradable

6. something able to naturally decay or break down
   - **b.** recycling

7. a rich soil that forms from decayed biodegradable material
   - **c.** compost

8. changing a material into something that can be used again
   - **d.** conservation

Understanding What You Learned

1. What are the four things most living things need?

2. What is the natural resource in which most food is grown?

3. When our garbage is picked up, where is most of it taken?

4. What might be damaged by dumping chemicals into the ground?

Applying What You Learned

1. Make a list of five things that are made from trees. Then, think of another natural resource that could replace the use of trees to make these same things.

2. Why are some of Earth's resources, like coal and petroleum, impossible to replace?

 3. Name three ways that you can help protect the environment.

For Your *Portfolio*

Think about the relationships among some living things. Draw a picture of a habitat that shows how several animals or plants depend on each other. Describe their relationships. Explain what might happen if a new plant or animal entered the habitat.

Changes in

Nature causes many changes in the environment. Some of nature's changes are slow. It may take years for wind to blow away a sandy beach. Some of nature's changes are fast. An erupting volcano quickly changes the environment. So does a forest fire sparked by lightning.

Human actions also cause changes to the environment. These changes are not natural changes. For example, a forest is cut down and birds have no place to nest. A dam is built to collect water for human use, flooding a valley.

Every change to an environment, no matter how small, can affect the whole environment.

The Big IDEA

When an environment changes, some organisms survive and reproduce; others die.

the Environment

CHAPTER SCIENCE INVESTIGATION

Learn more about endangered plants and animals. Find out how in your *Activity Journal.*

A51

Surviving Changes in the Environment

Find Out

- How a natural disaster affects living things in an environment
- How nature repairs itself after a disaster
- How humans help nature repair itself after a disaster
- How humans affect living things in an environment

Vocabulary

restore

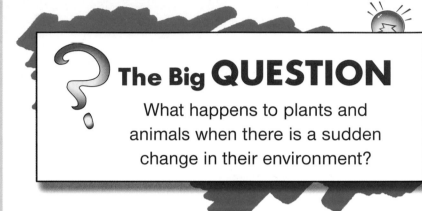

The Big QUESTION

What happens to plants and animals when there is a sudden change in their environment?

We are lucky to live in modern times. Scientists can often predict floods. They can tell us when a volcano is likely to erupt. They can warn us about tornadoes. Television and radio reports tell about fires and other disasters. These warnings help people avoid harm. Animals and plants don't get these same warnings. What happens to them when there is a sudden change?

Effects of a Natural Disaster

On May 18, 1980, in southwestern Washington, an earthquake shook the Mount Saint Helens volcano at 8:30 A.M. The quake caused the side of the volcano to collapse. Blasts caused by gases inside the volcano blew ashes, cinders, and gases out of the opening. At the same time, a huge column of ash blew upward into the air.

The force of the volcano tore millions of trees from the ground and laid them down like toothpicks. Six hundred square kilometers of forest land were destroyed. Sixty-one people near the volcano died. So did about 5000 deer, 200 black bears, 1500 elk, and many other animals. Two million birds and fish also died.

Scientists had thought that the volcano would erupt. They warned people to leave. Many people drove out of the area.

Mount Saint Helens after eruption

Of course, the animals could not drive away. But they may have sensed the trembling ground days before the blast. Some left the area. The animals that did not travel far enough away were killed. Those that traveled great distances from the volcano survived.

On the day of the eruption, an observer 22 km from the volcano saw something amazing. Right before Mount Saint Helens blew up, beavers living near the volcano slapped their tails on the water. Then they dove under. The beavers' actions signaled that a big change was coming. The beavers responded by taking cover.

Fireweed, lupine, and thimbleberry plants lived through the blasts. Of course, these plants could not move away. But their strong roots kept them anchored to the ground.

Nature Restores Itself

Nature always works to **restore**—or repair—itself after a disaster. Three weeks after the eruption of Mount Saint Helens, birds were flying near the volcano.

Seeds of trees and other plants destroyed during the eruption survived. When conditions were right for growth, these seeds began to grow.

Within a year after the eruption, many small animals had returned. Deer tracks marked the ash-covered ground. Herds of elk grazed on the grass that poked through the ash covering the ground.

Mount Saint Helens after restoration

Three years later, plants were growing in the area again. Insects, birds, fish, and other small creatures had also returned to their habitats.

Of course, there were not nearly as many of each kind of animal and plant as before. The eruption had killed a great many. But as time passed, their numbers steadily grew. When the Mount Saint Helens environment was safe again, the plants and animals there began to reproduce. As they reproduced year after year, their populations grew.

Humans Help Nature Restore Itself

Humans actively worked to help restore the area around Mount Saint Helens after the eruption. Workers from the U.S. Forestry Service planted 10 million fir tree seedlings. Other workers built new roads and trails. They built new visitors' centers to welcome campers and hikers.

There are still signs of the day the mountain erupted. It will take many years for tall forests to again cover the land. But someday, the entire landscape will be restored.

Worker planting fir tree seedlings

Many opossums and raccoons live in attics and find food in garbage cans, not forests.

Humans Cause Changes

Your environment didn't always look the way it does now, especially if you live in a town or city. Once it was the home of many different populations. Then things began to change. Trees were cut down to build houses and roads. Meadows became backyards, and swamps and ponds were filled in with soil.

Humans can cause sudden changes in an environment too. People sometimes make mistakes that harm other living things. Campers have left campfires

burning in the woods, starting forest fires. Accidental oil spills in the ocean pollute both the water and the shore, making life very hard for the plants and animals in the area.

When environments change very quickly, animals often do not have time to get to safety. Plants, of course, cannot move away from danger. They either survive where they are or they die. In time, nature can recover from many sudden changes. Other changes may change an environment forever.

Falcons nest on the rooftops of tall buildings in cities the same way they used to nest on mountain cliffs.

Rock dove with a nest in a sign

CHECKPOINT

1. How did the eruption of Mount Saint Helens affect the plants, animals, and humans living around it?

2. Describe some of the ways nature restores itself after a volcanic eruption.

3. What are some ways humans helped restore the area around Mount Saint Helens?

4. How do humans make it harder for plants and animals to live in their environment?

❓ What happens to plants and animals when there is a sudden change in their environment?

ACTIVITY

Cleaning Up Oil Spills

Find Out

Do this activity to see how hard it is to clean up an oil spill in the ocean.

Process Skills

Experimenting
Observing
Communicating
Measuring

WHAT YOU NEED

500-mL measuring cup

vegetable oil

clock or timer

Activity Journal

nylon mesh or cheesecloth

paper towels

blue food coloring

water

three shallow aluminum or glass pans

cotton balls

WHAT TO DO

1. **Measure** and pour 500 mL of water into each of the three pans. Add one or two drops of blue food coloring to the water in each pan.

2. Using the cap from the vegetable oil bottle, add exactly one capful of oil to the water in each pan.

3. Try to remove the oil from the water. **Experiment** by using a different material (cotton balls, paper towels, or mesh) to remove the oil from each pan. Use only one material to remove the oil from each of the pans. Set the timer and allow

yourself exactly four minutes to remove the oil from each of the pans.

4. **Observe** what happens as you try to remove the oil with each of the materials. **Record** your observations.

CONCLUSIONS

1. Were you able to remove any oil with the different materials? How can you tell?

2. Were you able to remove all of the oil with any of the materials? How can you tell?

3. How is this activity like cleaning up an oil spill in the ocean? How is it different?

ASKING NEW QUESTIONS

1. What might cause oil to spill in the ocean?

2. How might the spill affect fish, birds, and plants that live in or near the ocean?

3. Why is oil harder to clean up than water?

SCIENTIFIC METHODS SELF CHECK

✔ Did I **experiment** with different materials when trying to remove the oil from the water?

✔ Did I **observe** what happened?

✔ Did I **record** my observations?

Threats to Survival

Find Out

- What can be learned from dinosaur fossils
- What may have caused dinosaurs to become extinct
- Why other plants and animals become extinct

Vocabulary

fossils
adapt
extinct

The Big QUESTION

Why did some animals and plants become extinct?

*N*ot all kinds of plants and animals can adapt to change. Those that can't adapt disappear from Earth forever.

Dinosaur Fossils

About 215 million years ago, there were many types of dinosaurs. Some were only about the size of a chicken. Some were as tall as four-story buildings!

Large dinosaurs like this one once ruled Earth.

A60

Scientists have found fossils that show signs that dinosaurs lived many millions of years ago. **Fossils** are footprints, bones, or other remains of things that lived a long time ago. Much of what we know about early life on Earth is learned from studying these fossils.

By studying fossils, scientists can learn many things. They can tell where and when dinosaurs lived. They can tell what they probably ate and how they had their young. Dinosaur egg fossils have been found in Montana.

Dinosaurs lived on Earth for about 150 million years! To survive for so long, they had to adapt to their environments. To **adapt** is to change size, shape, or features over time to adjust to a changing environment. Dinosaurs were adapted in ways that helped them to eat the kinds of food they needed, to care for their young, and to protect themselves from other dinosaurs.

For example, those that had long necks could eat from treetops. Those that ate meat had sharp teeth. Most dinosaurs had thick skin that helped protect them from harm.

Dinosaur footprints

Dinosaur egg fossils

A61

Dinosaurs Disappeared

Something happened that made the dinosaurs' environment change too quickly for them to adapt. Different types of dinosaurs began to die. Over time, almost all of the different kinds of dinosaurs died. When every last one of a certain kind of plant or animal has died, its population has become **extinct** (ek stingt′). No one knows exactly why dinosaurs became extinct. Lizards alive today look like some of the dinosaurs that died long ago.

Some scientists believe a huge meteor, or rock from space, struck Earth. It hit Earth so hard that it caused a big dust storm that blocked out light from the sun. Without sunlight, plants died and so did dinosaurs.

Environments on Earth are constantly changing. Here's how one place may have changed.

Dinosaurs lived when it was warm and there was plenty of rain.

As it got colder, only organisms that could survive the ice and snow lived.

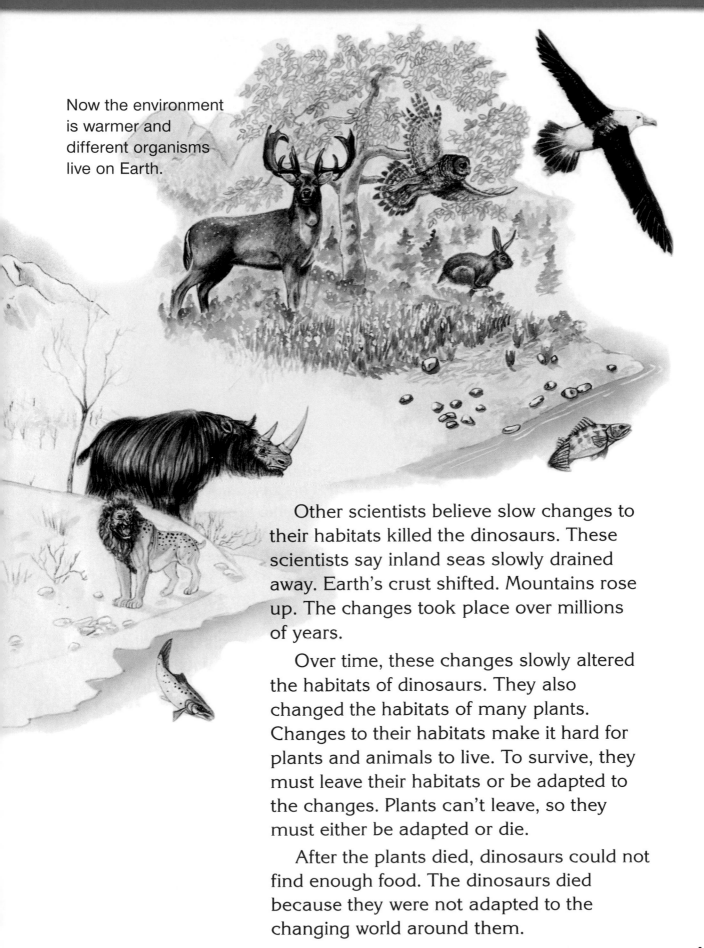

Now the environment is warmer and different organisms live on Earth.

Other scientists believe slow changes to their habitats killed the dinosaurs. These scientists say inland seas slowly drained away. Earth's crust shifted. Mountains rose up. The changes took place over millions of years.

Over time, these changes slowly altered the habitats of dinosaurs. They also changed the habitats of many plants. Changes to their habitats make it hard for plants and animals to live. To survive, they must leave their habitats or be adapted to the changes. Plants can't leave, so they must either be adapted or die.

After the plants died, dinosaurs could not find enough food. The dinosaurs died because they were not adapted to the changing world around them.

The dodo bird was about the size of a large turkey. It had short, stubby wings and could not fly. It lived on the island of Mauritius. Sailors killed the bird for food. Goats and pigs that sailors brought to the island ate the dodo's eggs and young. It became extinct in 1680.

The great auk became extinct in about 1850. Sailors visiting islands in the North Atlantic killed this bird for its fat, meat, and feathers.

Other Plants and Animals Became Extinct

Many kinds of animals and plants have become extinct as a result of natural changes. Changes in the weather killed some. Too many predators killed others. The food supply of many animals disappeared.

Just a few hundred years ago, there were three to five billion passenger pigeons living in beech and oak forests. Many of these trees were cut to build farms. The pigeons' habitat was gone. Then hunters killed the pigeons for food. The last passenger pigeon died in 1914.

CHECKPOINT

1. What information has been learned from dinosaur fossils?

2. What may have caused dinosaurs to become extinct?

3. Why did other plants and animals become extinct?

 Why did some animals and plants become extinct?

ACTIVITY

Drawing Animal Adaptations

Find Out

Do this activity to see how an organism that lived in the swamps with the dinosaurs might have adapted to change.

Process Skills

Communicating
Inferring

WHAT YOU NEED

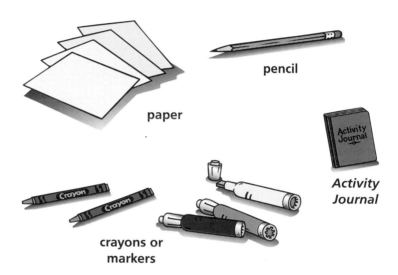

paper

pencil

crayons or markers

Activity Journal

WHAT TO DO

1. **Draw** a picture of a swamp creature.

2. **Draw** a second picture to show how its body changed when it moved to land after the swamps dried up. Label the picture "One Million Years Later."

3. **Draw** a third picture showing how it adapted after the climate changed. Label the picture "Two Million Years Later."

4. Under each picture, **explain** how the creature's changes helped it to be adapted to its environment.

5. **Infer** what kind of habitat the creature would need to survive.

6. **Record** your inference.

CONCLUSIONS

1. Share the pictures you drew with a classmate.

2. Ask your classmate to list the changes in each picture.

3. Ask your classmate to name other changes that might have helped your creature adapt over millions of years.

ASKING NEW QUESTIONS

1. What might cause your creature to make further changes?

2. What would life on Earth be like today if dinosaurs *had* been able to adapt to change?

SCIENTIFIC METHODS SELF CHECK

✔ Did I **communicate** my ideas about the swamp creature by drawing pictures?

✔ Did I **infer** from the drawings what kind of habitat the creature would need to survive?

✔ Did I **record** my ideas?

Survival in a Changing World

Find Out

- How living things change over time
- What it means to be endangered
- How rain forests are important
- How life on Earth depends on a balance of living things

Vocabulary

endangered

The Big QUESTION

Why are some plants and animals endangered?

What happens when organisms disappear? When organisms no longer exist on Earth, they have become extinct. Sometimes extinction is a natural change. It has been happening for millions of years. In fact, most living things that once existed are now extinct! Sometimes human actions cause certain kinds of plants and animals to become extinct.

Change Over Time

Suppose you traveled back in time thousands or even millions of years. You wouldn't see the same plants and animals you see today. You might see dinosaurs, woolly mammoths, or saber-toothed cats. You would see huge ferns and other now-extinct plants.

Due to natural changes, these and many other animals and plants no longer exist. They became extinct because their environments changed too much. These animals and plants could not adapt.

Other animals and plants did adapt to changing environments. For example, wild horses were once very small and had four toes on their feet. Over millions of years, horses' bodies slowly changed. They became larger and their feet changed shape over time. Horses today are large animals with hard, single-toed feet called hooves.

Scientists do not know exactly how these adaptations helped horses to survive. One theory is that solid, single-toed feet helped horses to run more quickly than other animals. This may have allowed them to escape from predators more easily. There may be many different reasons why horses changed over time. Whatever the reason, we know that horses survived because they adapted over time to their environment.

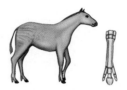

Eohippus
55 million years ago (four-toed)

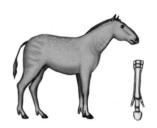

Mesohippus
40 million years ago (three-toed)

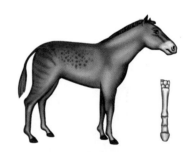

Merychippus
25 million years ago (three-toed)

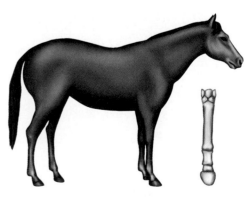

Pliohippus
5 million years ago (one-toed)

After millions of years of change, horses now have single-toed feet called *hooves.*

Equus
2 million years ago (one-toed)

A69

Endangered Plants and Animals

Living things in danger of becoming extinct are **endangered.** There are thousands of kinds of plants and animals that are endangered.

Living things that are endangered become so few in number that they may not be able to have young. If this happens, their whole population will become extinct.

There are many reasons to be concerned about endangered organisms. When an endangered animal or plant becomes extinct, the food chain is affected. Other living things that depend on the animal or plant could also become extinct.

Today, many kinds of whales are endangered. Whales have been hunted for hundreds of years. Parts of whales are used for many things. Some types of glue, perfume, soap, and cattle feed are made from whale parts.

A large number of other living things are also endangered. Among them are the rhinoceros, a bird called the California condor, and the African elephant. Endangered plants include the pitcher plant, the hedgehog cactus, and the western lily.

Rhinoceros

California condor

Western lily

One way to protect these and other endangered plants and animals is to make laws. In 1973, the U.S. Congress passed the Endangered Species Act. This law makes it illegal to hunt, collect, or harm any endangered plant or animal.

This was not the first law to protect endangered plants and animals. Laws meant to protect wildlife were first made in the year 1900. Still, many kinds of animals have become endangered or extinct.

Because their jobs were affected, some people were against the Endangered Species Act. For example, the act kept lumber companies in the Pacific Northwest from cutting down trees in certain forested areas. The areas were home to the spotted owl, an endangered bird. Supporters of the act said each pair of owls needed large areas of forest for their habitat. The lumber companies strongly disagreed.

People continue to disagree about how to best protect endangered plants and animals in the United States. The laws often cause disagreements. The laws that are not changed by the courts must be obeyed.

Spotted owl

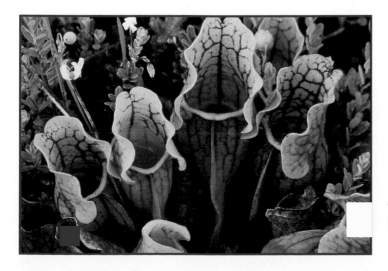

Pitcher plant

Rain Forests in Danger

Our laws cannot protect living things in other parts of the world. A million kinds of plants and animals in tropical rain forests are endangered or already extinct!

Every day, people cut down rain forests in many different parts of the world. Some people cut down rain forest trees so that they can sell their wood as lumber. Some people clear away all the trees and other plants from rain forests so that they can use the land to grow crops or graze animals. When the rain forests' trees are cut down and taken away, a very important part of the ecosystem is lost. This destroys the habitats of all living things in the area.

If the rain forests disappear, these plants and animals will become extinct. Their populations will die because they will have no habitat in which to live.

Rain forests are important parts of our global ecosystem. Trees and plants supply much of Earth's oxygen. Many kinds of rain forest plants are also used to make medicine for human use. Scientists believe that some plants growing in the rain forests could be used to make important new kinds of medications. If these plants are destroyed before scientists can find them, new medicines that could help people may never be discovered.

Rain forests are home to hundreds of thousands of different kinds of organisms. Shown here are the cotton-top tamarin, the poison-dart frog, and the scarlet macaw.

Our Global Habitat

Although there are many different types of ecosystems on our planet, each one of them is part of the large ecosystem that is Earth. Every part of our global ecosystem depends on the other parts.

Many people believe that our global ecosystem is in danger. Have you ever heard of the "greenhouse effect"? The greenhouse effect is something that some scientists think may be happening to our planet. As people burn fuel to power cars, homes, and factories, gases build up in the air around Earth. These gases could trap heat in our environment, causing our planet's temperature to rise slowly over time. Many people believe that the greenhouse effect may cause big changes in our global ecosystem. These changes could make life hard for many plants and animals.

We must protect our entire global ecosystem. Why? Because maintaining the balance of living things on Earth is key to the survival of plants, animals, and humans.

The bird-of-paradise grows in tropical rain forests.

CHECKPOINT

1. Describe how horses changed over time.
2. Explain what it means to be an endangered plant or animal.
3. Why are the rain forests so important to Earth's ecosystem?
4. Why must we protect our ecosystem?

 Why are some plants and animals endangered?

A73

ACTIVITY

Investigating What Happens in a Greenhouse

Find Out

Do this activity to see how a greenhouse traps heat.

Process Skills

Constructing Models
Observing
Measuring
Communicating

WHAT YOU NEED

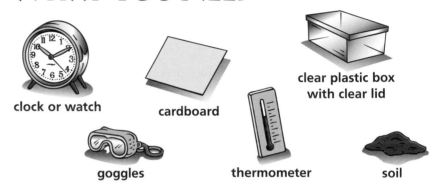

clock or watch

cardboard

clear plastic box with clear lid

goggles

thermometer

soil

Safety! *Wear safety goggles at all times during this experiment.*

WHAT TO DO

1. Spread 3 cm of soil in the plastic box.

2. Place a piece of cardboard into the soil in the middle of the box.

3. Prop the thermometer on the cardboard. The bulb of the thermometer should be facing up and above the soil. **Read** the thermometer and **record** the temperature in the box.

4. Put the box where it will receive direct sunlight.

   Safety! *Be very careful when handling thermometers. If a thermometer breaks,* do not touch the pieces. *Tell a teacher or other adult right away.*

5. Wait 10 minutes. **Read** the thermometer and **record** the temperature.

6. Move the box out of the direct sunlight. Wait for the box to return to the original room temperature. Then, place the lid on the box and return the box to the sunlight.

7. Wait 10 minutes. **Read** the thermometer and **record** the temperature.

8. Repeat Steps 4–7 and **record** each temperature reading.

CONCLUSIONS

1. Compare the air temperatures in the plastic boxes.

2. Based on your measurements, which plastic box is like a greenhouse?

3. How is your model like the greenhouse effect on Earth? How is it different?

4. Were your measurements the same the second time you made them?

ASKING NEW QUESTIONS

1. What would happen if you used water in the plastic boxes instead of soil? Develop a plan to answer this question. Test your answer.

SCIENTIFIC METHODS SELF CHECK

✔ Did I follow the directions to **construct a model** of a greenhouse?

✔ Did I **record** and **measure** carefully?

✔ Did I **observe** the changes in temperature between the two boxes?

CHAPTER 3

Review

Reviewing Vocabulary and Concepts

Write the letter of the word that completes each sentence.

1. Evidence of plants and animals long extinct is a ___.
   - **a.** species
   - **b.** scavenger
   - **c.** fossil
   - **d.** habitat

2. Plants and animals that have disappeared entirely from Earth are ___.
   - **a.** extinct
   - **b.** scavengers
   - **c.** endangered
   - **d.** threatened

3. Living things that exist in so few numbers that they are in danger of disappearing are said to be ___.
   - **a.** communities
   - **b.** extinct
   - **c.** dinosaurs
   - **d.** endangered

4. Huge tropical areas that are home to hundreds of thousands of living things are ___.
   - **a.** communities
   - **b.** populations
   - **c.** rain forests
   - **d.** greenhouses

Match the definition on the left with the correct term.

5. what nature does to bring back plants and animals after a volcano   **a.** adapt

6. what plants and animals do to survive changes in the environment over time   **b.** medical research

7. a reason to preserve Earth's rain forests   **c.** habitat

8. an environment in which a plant or animal lives   **d.** restore

Understanding What You Learned

1. How did animals' numbers grow at Mount Saint Helens after the volcano erupted?

2. List three reasons that plants or animals become extinct.

3. List two ways in which horses have changed over time.

4. Why might the disappearance of one kind of plant or animal from an environment threaten others?

Applying What You Learned

1. How were dinosaurs adapted to their environment?

2. Write down two ways in which actions of humans can contribute to the extinction of plants or animals.

 3. Explain what happens to organisms when an environment changes over time.

For Your **Portfolio**

Think about what it means for an organism to be extinct. Then list five plants or animals and describe what changes in their environments might force them to adapt to the change, leave the environment, or die.

Unit Review

Concept Review

1. Describe how an animal or a plant that you are familiar with gets what it needs from its environment.

2. Why should everyone work to take care of Earth?

3. Would you expect your community to have the same populations 100 years from now as it does today?

Problem Solving

1. Describe a food chain in your neighborhood that you know about.

2. What rules could you write to protect your school environment?

3. What would happen to the living things in an empty field if many houses were built there?

Something to Do

With a group, create a mural that shows an ecosystem, such as a forest, as it appears today. Around the mural include smaller posters that show blowups of habitats within the ecosystem, such as a stream, a nest in a tree, and the space under a rock. A paper arrow leading from each poster to the mural will show where the habitat fits into the ecosystem. Then create another mural showing the same area as you think it might appear a long time from now.

UNIT B

Earth Science

1 Earth and Space

What do the sun, the moon, and Earth have in common? They are all part of a system in space called the solar system. The sun is at the center of our solar system.

The planet Earth is our home. Earth, along with everything else that revolves around the sun, is part of our solar system.

Investigating the solar system can be fascinating. Take a closer look.

The Big IDEA

Earth is part of our solar system.

CHAPTER SCIENCE INVESTIGATION

Learn whether there is a pattern in what we see of the sun and the moon. Find out how in your *Activity Journal.*

Earth, Sun, and Moon

Find Out

- How shadows are created
- What happens during an eclipse
- What the phases of the moon are

Vocabulary

shadow
opaque
eclipse
phases

The Big QUESTION

Why does the moon appear to change?

Have you ever made shadow puppets on the wall or a movie screen? Have you ever stopped for a rest in the cool shade of a tree? Do you ever put your hand up to shade your eyes from the bright sun? If you have, then you know something about shadows. Earth and the moon make shadows too. Did you know that sometimes you can see these shadows?

Sunlight and Shadows

A **shadow** is a dark area caused by the blockage of light. Three things are needed to form a shadow: a source of light, an object to block the light, and a background for the shadow to fall onto. The shadow forms when the object comes between the light source and the background. The shadow is the dark shape you see on the background.

Think about the last time you made shadow puppets. The light source may have been a lightbulb or movie projector. Your hand was the object that blocked the light, and the shadow fell upon the wall or movie screen. The shade of a tree is caused by the tree blocking the light from the sun. The shadow of the tree is cast on the ground. When you shade your eyes with your hand, your hand blocks the light and the shadow falls on your face.

All opaque objects cast shadows. An **opaque** (ō pāk′) object does not let light pass through it. If you can't see through an object, it is opaque and it will block light. Transparent objects, such as glass windows and some types of plastics, allow light to shine through, so they do not cast shadows onto other objects.

The shape of a shadow depends on the shape of the object blocking the light. The size of a shadow can depend on how large an object is or how close it is to a light source. Larger objects create larger shadows. The closer an object is to a light source, the larger its shadow will be.

Making shadow puppets

The larger the object is, the larger the shadow will be.

During a solar eclipse the moon's shadow falls on Earth. While orbiting Earth, the moon moves between the sun and Earth, blocking the sun's light and casting a shadow on Earth for several minutes.

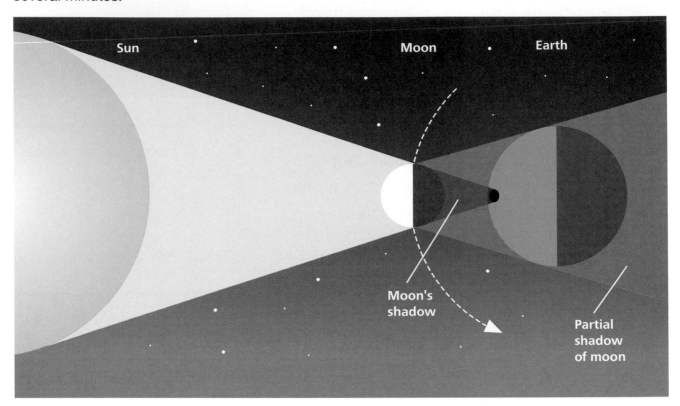

Sun　　Moon　　Earth

Moon's shadow

Partial shadow of moon

Eclipses

Both the moon and Earth always cast a shadow. Usually we can't see their shadows because the shadows are cast out into the darkness of space. The only time we can see the moon's or Earth's shadow is during an eclipse. An **eclipse** (i klips′) happens when one object passes into the shadow of another object. The sun is the source of light in shadows of the moon and Earth.

Every year lunar and solar eclipses can be seen somewhere on Earth. A lunar eclipse lasts an hour or two. People on the side of Earth facing the moon have a chance to see a lunar eclipse as the moon moves into Earth's shadow.

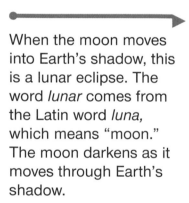

When the moon moves into Earth's shadow, this is a lunar eclipse. The word *lunar* comes from the Latin word *luna,* which means "moon." The moon darkens as it moves through Earth's shadow.

8. From night to night, you can watch the moon become a crescent again.

9. After $29\frac{1}{2}$ days, the moon is back between Earth and the sun. It's time for another new moon.

1. The phase when the moon is between Earth and the sun is called a new moon. You can't see the new moon because all of the side lighted by the sun faces away from Earth. There is no moonlight on those nights.

As seen from Earth

2. After three or four days, the moon has moved in its orbit so that you can see a thin sliver of the lighted side. This is a crescent (kres′ ent) moon.

3. After seven or eight days, the moon has moved one-fourth of the way around Earth. It is called a first quarter moon.

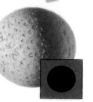

Phases of the Moon

Although the moon is often the brightest light we can see at night, it really makes no light of its own. What we call moonlight is really sunlight reflected off of the moon. When we say that one part of the moon is "lighted," we are talking about the part that reflects the sun's light.

The sun is always shining on part of the moon, just as it is always shining on part of Earth. The same side of the moon always

7. After 21 or 22 days, the moon has traveled three-fourths of the way around Earth. It only has one-fourth of the way to go, so it is called a last quarter moon.

6. The moon keeps moving. You see less of its lighted side.

5. After 14 or 15 days, the moon is halfway around its orbit. It is on the opposite side of Earth from the sun. Now you can see the entire lighted side of the moon. This phase is called a full moon.

4. As the moon orbits Earth, you can see more and more of its lighted side.

faces Earth. How much reflected light you see depends on where the moon is in its orbit around Earth. The different amounts of reflected light you see are called **phases** of the moon.

CHECKPOINT

1. List the three things needed to create a shadow. Explain how a shadow is made.

2. What happens during an eclipse?

3. Describe what we see during the phases of the moon and why we see what we do.

 Why does the moon appear to change?

ACTIVITY

Modeling the Moon's Phases

Find Out

Do this activity to show why the moon appears to change shape during its monthly phases.

Process Skills

Constructing Models
Observing
Communicating

WHAT YOU NEED

two pieces of white construction paper

toothpick

scissors

masking tape

one piece of yellow construction paper

small foam ball

Activity Journal

black marker

WHAT TO DO

1. Make a model of the sun by cutting a large circle from the yellow construction paper. Tape the sun to the middle of the front wall in your classroom.

2. Cut each of the pieces of white construction paper into four equal squares. Label each square with a capital letter from A to H. Put "A" on the first square, "B" on the second square, and so on. Tape square A right under the sun. Tape squares B–H around the room as shown.

3. Use a marker to color half of the foam ball black. In this activity, you will be Earth. The black-and-white ball will be the moon. Stick a toothpick into the moon anywhere that black and white meet. Use the toothpick as a handle.

4. Have your partner move the moon around Earth (you), always keeping the white side of the moon facing the sun. **Observe** the moon carefully as it moves around you, turning as necessary to keep the moon in sight. **Observe** the shape of the white part of the moon as it passes each piece of white paper.

5. Repeat Step 4, changing places so that you are the moon and your partner is Earth. **Record** and compare your observations.

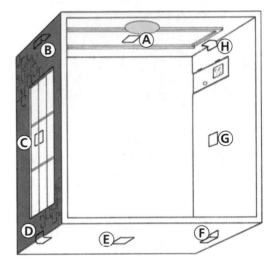

CONCLUSIONS

1. Did the moon really change shape? Why or why not?

ASKING NEW QUESTIONS

1. Why are we able to see only part of the moon at certain times?

SCIENTIFIC METHODS SELF CHECK

✔ Did I **make a model** of the moon's phases?

✔ Did I **observe** my model moon as it moved around Earth?

✔ Did I **record** my observations?

The Nine Planets

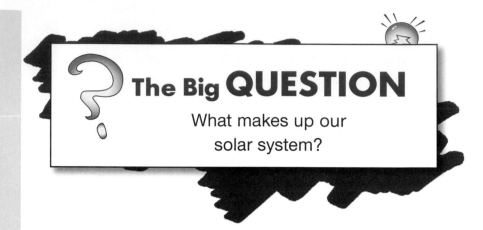

The Big QUESTION

What makes up our solar system?

Imagine you are sitting in your backyard gazing at the stars. All of a sudden there is a flash of light and a flying saucer drops from the sky! The spacecraft opens and out marches a group of little green aliens! One of the aliens turns to the others and says, "We have landed on Earth, the third rock from the sun."

"Mom!" you call out. "The aliens from space are here!" It's a neat dream, but as far as we know, it has never happened. However, we do know a lot about what is out there and what's going on in the solar system.

Center of the Solar System

The sun is a star. It is the closest star to Earth, which is why it looks so much larger than the other stars we see.

The sun is a big ball of very hot gases. Without the sun, we would not be able to live on Earth. It gives off energy that is converted to heat and light.

The sun is the center of the **solar system.** The solar system is made up of everything that travels around the sun, including planets, moons, comets, meteoroids, and asteroids. There are nine known planets in the solar system. A **planet** is a heavenly body that moves around a star. Each of the planets travels around the sun in a path called an **orbit.**

Because Earth turns as it orbits the sun, the sun seems to move across the sky during the day. Earth turns, or rotates, one full circle every day and completes one orbit around the sun every year. The position of the sun in the sky changes from hour to hour during the day, and from season to season during the year.

Orbits of the planets closest to the sun

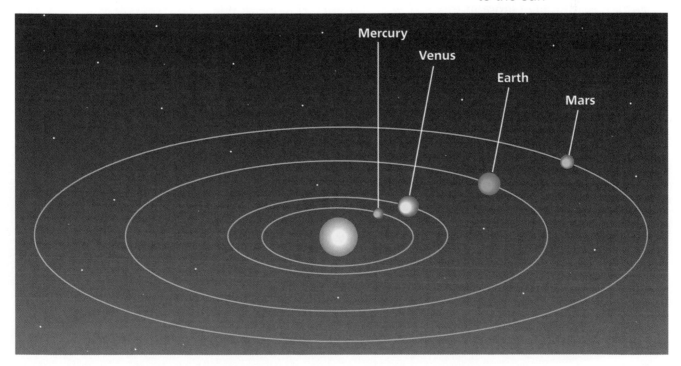

Mercury
Venus
Earth
Mars

The Inner Planets

Each of the nine planets has its own place in our solar system. You've probably guessed from our story where Earth's place is. Let's look at each planet as we take an imaginary trip away from the sun.

The four planets closest to the sun are called the inner planets. The inner planets are smaller and warmer than most of the outer planets. The inner planets are made mostly of rock.

Mercury is the closest planet to the sun. It is hot and dry. Mercury is the second smallest of the nine known planets. Its surface is covered with craters.

Earth is the third planet in the solar system. It is the only planet in the solar system where life is known to exist. It is the only planet known to have liquid water. Earth is the largest of the four inner planets.

Venus is the second planet from the sun. It is the hottest planet because it is surrounded by thick clouds that trap heat from the sun. You can often see Venus shining brightly at sunrise or sunset. It is sometimes called the morning or evening star even though it is not a star.

Mars is the fourth planet from the sun. It is more like Earth than any other planet. It has four seasons, ice caps at its poles, and traces of water vapor in its atmosphere.

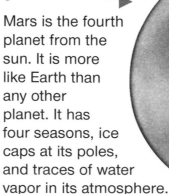

The Outer Planets

There are five known outer planets in our solar system. Most of them are much larger and colder than the inner planets.

Saturn is the sixth planet from the sun. It is almost as big as Jupiter and is also made mostly of gases. Saturn is best known for its rings. It has the most moons of any planet.

Jupiter is the fifth planet from the sun. It is the largest planet in our solar system. Jupiter is made mostly of gases. It has a set of rings, but they are thin and hard to see. Jupiter is best known for its giant red spot, a storm three times larger than Earth.

Uranus is the seventh planet in our solar system. Uranus is also made mostly of gases. It is about four times the size of Earth. It has rings that are too thin to see with our eyes alone. Uranus is very cold.

Neptune, the eighth planet, is similar to Uranus. It also has rings that are too thin to see and is very cold. Neptune is made of gases. Neptune, Jupiter, Saturn, and Uranus are often called the Gas Giants.

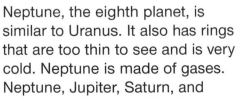

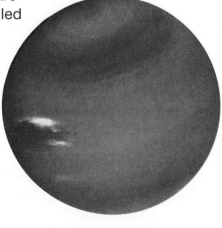

Pluto is the ninth planet in our solar system. It is also the smallest and coldest. It is made mostly of ice and rock. Pluto has an unusual orbit. Part of its orbit crosses inside Neptune's orbit. When that happens, it becomes the eighth planet from the sun. Pluto was closer to the sun than Neptune from 1979 to 1999.

Telescopes

We can see many things in space with our naked eye—if we know where to look. We can see the moon, many stars, and even other planets. Sometimes, though, people want to view things that are too far away to see. That's when telescopes are helpful.

A **telescope** is an instrument used to observe objects that are very far away. A telescope is a tube with a special kind of glass at each end called a lens. When you look at something through a telescope, it seems closer and brighter.

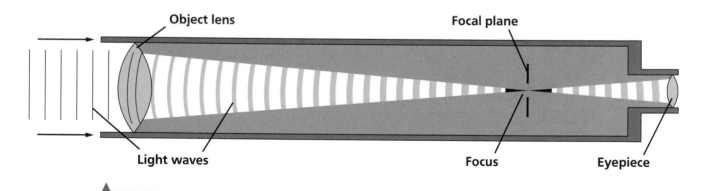

Object lens **Focal plane**

Light waves **Focus** **Eyepiece**

Refracting telescopes have a large lens at the front of the tube to gather light. The large lens is called the object lens. The light rays pass through the object lens and are bent so that they come together at the end of the tube. The light rays are directed into a smaller lens called the eyepiece. The eyepiece spreads the light and makes the object you are looking at appear bigger.

Reflecting telescopes use mirrors instead of an object lens to gather light. The light is gathered through a curved mirror at one end of the telescope. The light is then reflected back to a smaller mirror and finally to a side opening where the lens is located.

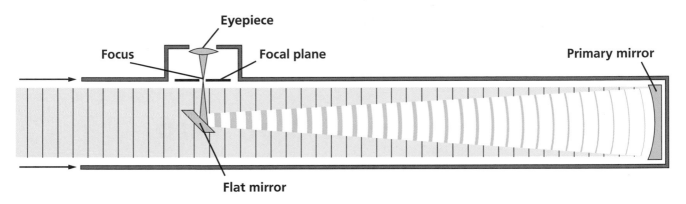

Focus · Eyepiece · Focal plane · Primary mirror · Flat mirror

No one knows who made the first telescope, but many historians give credit to a man named Hans Lippershey. Lippershey was an eyeglass maker. He is thought to have made the first telescope in the early 1600s.

Two main types of telescopes are the refracting telescope and the reflecting telescope. Both kinds work by collecting light from an object and then using that light to make a tiny picture called an image.

CHECKPOINT

1. What is an orbit?
2. What do most of the inner planets have in common?
3. What do most of the outer planets have in common?
4. How does a telescope work?
 What makes up our solar system?

ACTIVITY

Making a Model Solar System

Find Out

Do this activity to see how a model of the solar system can help you compare the distances of the planets from the sun and from each other.

Process Skills

Constructing Models
Measuring
Using Numbers

WHAT YOU NEED

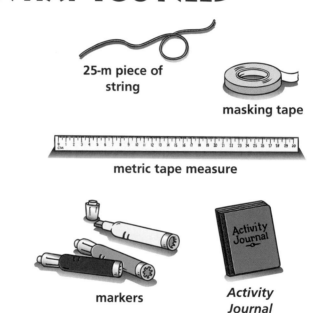

25-m piece of string

masking tape

metric tape measure

markers

Activity Journal

WHAT TO DO

1. Work with a partner. Mark the string at one end with a piece of masking tape. Write "Sun" on the tape.

2. **Measure** 25-cm from the sun along the string. Mark this point with tape, and write "Mercury" on the tape.

3. Measure 46 cm from the sun along the string. Mark this point with tape, and write "Venus" on the tape.

4. Using the following information, keep measuring along the string and using the tape to label the planets. Make each measurement from the sun, not from another planet.

Earth—63 cm
Mars—97 cm
Jupiter—3.30 m
Saturn—6.05 m

Uranus—12.16 m
Neptune—19.05 m
Pluto—25 m

5. Lay the string out straight with you at the sun end and a partner at the Pluto end. Have another person stand at Earth.

6. Using a colored marker, color the tape of the inner planets. Use a different color for the outer planets.

CONCLUSIONS

1. Which planet is about halfway between the sun and Pluto?

2. Compared to the outer planets, how far from the sun is Earth?

ASKING NEW QUESTIONS

1. Which planet goes around the sun in the shortest time?

2. Which planet takes the longest time to orbit the sun?

SCIENTIFIC METHODS SELF CHECK

✔ Did I carefully **measure** the distance of each planet from the sun in my **model?**

✔ Did I **use numbers** accurately when measuring distances in my model?

Constellations

The Big QUESTION

Why do constellations appear to change with the seasons?

Star light, star bright, first star I see tonight ...There is something fascinating about looking up at a sky full of stars. What are stars made of? Where did they come from? Why do we see different groups of stars at different times of the year?

The Milky Way

If you look up at the sky on a clear night, you may see thousands of stars. All of those stars, along with many more that you can't see, make up the **galaxy** we live in. Our galaxy is called the Milky Way.

The sun is one of the stars in the Milky Way. That means that our solar system is inside the Milky Way galaxy.

We know that the sun is made of hot gases. So are the other stars in the Milky Way. **Stars** are huge balls of hot gases. The two main gases that make up stars are hydrogen and helium.

Stars are formed from huge clouds of gas and dust. The gases and dust swirl together and collapse to form balls that become stars.

Long, long ago, people noticed that stars formed patterns in the sky and that those patterns did not change. These stars reminded some people of pictures, much like we might imagine seeing pictures in clouds. They called the patterns of stars **constellations** (kon′ stəll lā′ shəns) and gave them names. Some constellations even have stories to go with them.

Two Views of the Milky Way Galaxy

Position of our solar system

Position of our solar system

Our solar system is located in the Milky Way galaxy.

Our solar system is part of an even larger system called the **Milky Way Galaxy** (gal′ ek sē). There are billions of stars in it. Our sun is one of them.

The sun, Earth, and the moon are part of a larger system called the **solar system**. The sun is its center. There are other planets besides Earth. Do you know any of their names? Some of the planets have moons. Besides the sun, the planets, and their moons, there are other large and small objects in the solar system. You may know them as comets (kom′ itz); asteroids (as′ tə roidz′), and space dust.

Everything in the solar system orbits the sun. The sun's gravity is so strong it holds the whole solar system in place. The sun is also the source of light. Without it, the solar system would have almost no light and little heat.

B21

The Big Dipper

The constellation that you are probably most familiar with is the Big Dipper, a pattern of stars that looks like a water dipper or ladle.

Have you ever noticed that stars seem to rise and set just like our sun? This is because Earth rotates. As Earth completes one full rotation every 24 hours, the stars appear to rise, move across the night sky, and set. The stars you see depend upon your location on Earth and the time of year.

The Big Dipper is easy to find and is a useful constellation. Two stars in the Big Dipper's bowl point to the North Star. If you learn to find the North Star, Polaris, you will always know which direction is north.

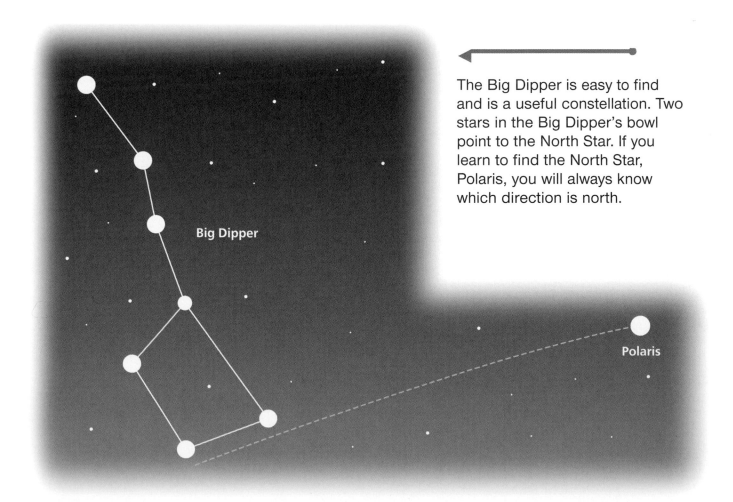

Big Dipper

Polaris

If you look at the same constellation at the same time of night about two weeks apart, you will notice that it is in a slightly different place the second time you look. The stars seem to move because Earth is moving. Because Earth orbits the sun, some constellations gradually seem to move out of sight, and new ones come into view. That is why we say we have "winter constellations" and "summer constellations."

The Big Dipper is always visible in the northern hemisphere, but it is not always in the same position. The position in which we see it changes as Earth revolves around the sun.

The Big Dipper is part of a larger constellation called the Great Bear. Can you find the Big Dipper inside this constellation?

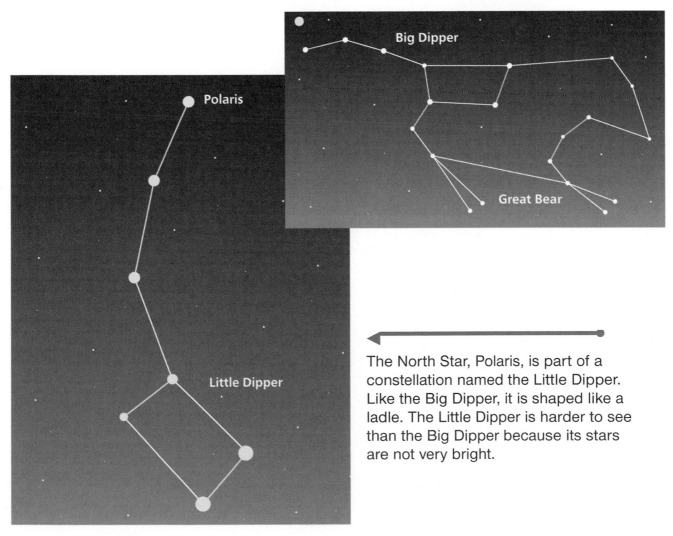

Big Dipper

Great Bear

Polaris

Little Dipper

The North Star, Polaris, is part of a constellation named the Little Dipper. Like the Big Dipper, it is shaped like a ladle. The Little Dipper is harder to see than the Big Dipper because its stars are not very bright.

Mapping the Sky

If you are interested in stargazing, some items may be helpful to you. You can look at a star chart. A star chart is like a map of the sky. It will show you where to look in the night sky to find the constellations you can see during each season of the year.

This star chart shows a view of the night sky as it looks from the north pole. Polaris, the North Star, is in the center of the chart to show that it is directly overhead at the north pole.

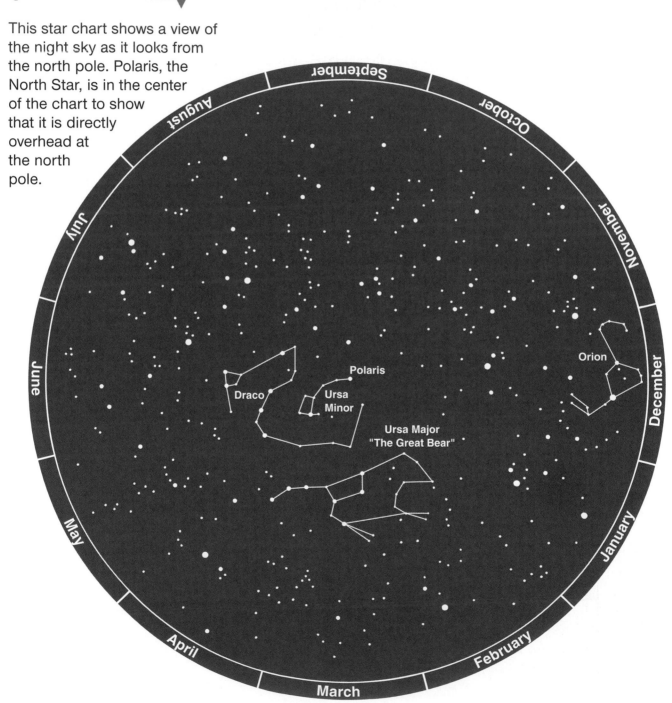

Look between the Big Dipper and the Little Dipper and you may see a dragon! Draco, the dragon, has a head made of four stars.

You may also want to visit a planetarium. A **planetarium** is like a star theater. In a planetarium, machines are used to project the images of stars onto a dome-shaped screen. The machines show how the constellations rise and set and change their positions with the seasons. Planetariums are great places to learn about the solar system, the galaxy, and the universe.

Orion, the hunter, is easy to recognize because he has two very bright stars in opposite corners. Do you think this constellation looks like a hunter?

CHECKPOINT

1. What is the Milky Way?
2. Why is the Big Dipper a useful constellation?
3. How is a star chart useful for finding constellations?

 Why do constellations appear to change with the seasons?

ACTIVITY

Making a Shoe Box Constellation

Find Out

Do this activity to see how to make a model of a constellation.

Process Skills

Constructing Models
Measuring
Observing
Communicating

WHAT YOU NEED

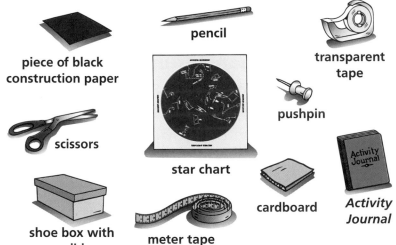

piece of black construction paper

pencil

transparent tape

scissors

star chart

pushpin

shoe box with a lid

meter tape

cardboard

Activity Journal

WHAT TO DO

1. **Measure** and then cut a 1-cm hole in one end of the box and a 17-cm × 13-cm hole in the opposite end.

2. Choose a constellation on the star chart. Place the construction paper and cardboard under the star chart and carefully use the pushpin to poke a hole through each star in your constellation.

3. Tape the construction paper over the large opening so that the holes are toward the center of the opening. Put the lid on the box.

4. Hold the box up to a bright light and **observe** through the small hole. **Record** your observations.

5. Trade boxes with your classmates, and find their constellations on your star chart.

6. **Draw** and **label** your constellation.

CONCLUSIONS

1. Which constellation did you pick?

2. Name some of the constellations that your classmates picked.

ASKING NEW QUESTIONS

1. How is the light from the stars different from the light from the moon?

2. Why don't we see the constellations during the daytime?

SCIENTIFIC METHODS SELF CHECK

✔ Did I **measure** the holes that I cut in the ends of the box to make sure they were the right size?

✔ Did I **observe** my constellation **model** and **record** my observations?

Review

Reviewing Vocabulary and Concepts

Write the letter of the word or phrase that completes each sentence.

1. Everything traveling around the sun is part of the ___.
   a. solar system   **b.** outer planets
   c. inner planets   **d.** star system

2. The paths planets travel around the sun are ___.
   a. phases   **b.** shadows
   c. circles   **d.** orbits

3. The Milky Way is what we call Earth's ___.
   a. orbit   **b.** eclipse
   c. galaxy   **d.** star

4. A ___ forms from a huge cloud of very hot gases.
   a. meteoroid   **b.** star
   c. planetarium   **d.** comet

Match the definition on the left with the correct term.

5. An object that does not let light pass through it   **a.** eclipse

6. A heavenly body that moves around a star   **b.** opaque

7. The name for patterns of stars in the night sky   **c.** constellations

8. What happens when one object passes into the shadow of another object   **d.** planet

Understanding What You Learned

1. What are the three things needed to make a shadow?

2. Give three examples of opaque objects.

3. List the moon phases that follow the new moon.

4. What are the inner planets?

5. Name two things about the inner planets that make them different from the outer planets.

Applying What You Learned

1. What happens during an eclipse?

2. Why is the North Star important?

3. In what way is the Big Dipper helpful?

4. Why does the Big Dipper seem to change position in different seasons of the year?

 5. Explain what the planets and other heavenly bodies within the solar system have in common.

For Your **Portfolio**

Select one of the planets visible in the night sky (Mars, Venus, or Saturn) and one constellation that appears in our sky throughout the year (Cassiopeia, Cepheus, Draco the Dragon, Orion, or the Big or Little Dipper). Make a sky diary. Try to see your planet and constellation each night for a month. Record what you see in your diary, even if it is cloudy and you cannot see the night sky.

Earth and Its

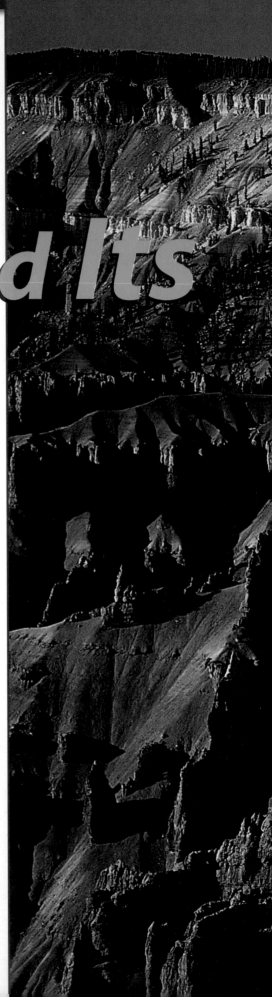

Have you ever played in the dirt? Did you ever wonder how far down you could dig? You can find soft soil, hard soil, rocks, and clay. Under the surface of the Earth are hot rocks and molten, or liquid, rock.

Earth has many layers. On the surface, the ground can be flat and low or steep and high. It can dip down into a valley. It can be a small hill or a great, tall mountain. Sometimes the surface is so high that it appears to reach the clouds.

Earth is always changing. Rocks are always being formed. Mountains are always changing, little by little. We can't always see these changes right away, but they are happening.

The Big IDEA

Earth has internal layers and a surface that changes over time.

Many Layers

SCIENCE INVESTIGATION

CHAPTER

Investigate how fossils are formed. Find out how in your *Activity Journal.*

Earth's Composition

Find Out

- What Earth's layers are
- How scientists study Earth

Vocabulary

crust
mantle
core

The Big QUESTION

What is inside Earth?

You walk on it. You run on it. Your school is built on it. But have you ever thought about what it is you're living on?

We live on the surface of Earth. The rocks, soil, and water we see are just the top layer of planet Earth. There are about 6400 km between the surface and the center of Earth.

Earth is like an onion. When an onion is cut in half, many layers can be seen. Earth also has many layers. The layers underneath affect what happens on the surface.

Different Layers of Earth

The Crust

The **crust** is the upper layer of Earth. The crust is thicker under the continents than it is under the oceans.

The crust is made up of different kinds of rocks. Rocks have their own properties like hardness, color, and shininess. Rocks are made, broken down, and remade all over again. Rocks near Earth's surface are being changed by water, weather, pressure, and living things. Deep inside Earth, tremendous heat and pressure change rocks too.

New layers of Earth are added to the crust all the time. This happens as all the different kinds of rocks are made.

In many places, the upper layer of Earth is covered with soil. Soil is made up of broken down rocks, minerals, and plant and animal matter. The upper layer of Earth can also contain fossils, which are usually found in rocks.

We live on the surface of Earth. Underneath us are many layers of different kinds of rocks. Some of these layers can be more than 40 km thick.

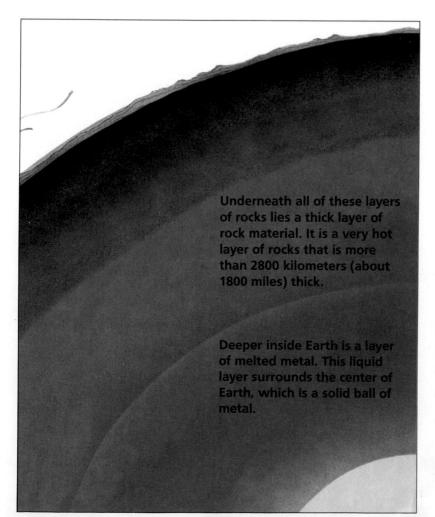

Underneath all of these layers of rocks lies a thick layer of rock material. It is a very hot layer of rocks that is more than 2800 kilometers (about 1800 miles) thick.

Deeper inside Earth is a layer of melted metal. This liquid layer surrounds the center of Earth, which is a solid ball of metal.

If we could go deeper into Earth, we would find more layers of rock. These rock layers are very hot. These layers circle around the center of Earth, which is a solid ball of metal.

B33

The Mantle and the Core

The **mantle** (man′ təl) is the layer of Earth that is just underneath the crust. It is made mostly of rock. The upper mantle has two areas, or zones. The lower zone is soft, and the upper zone is rigid, or hard. The hard part can move over the softer, lower part. The crust of Earth is divided into hard plates that move slowly on the soft part of the mantle.

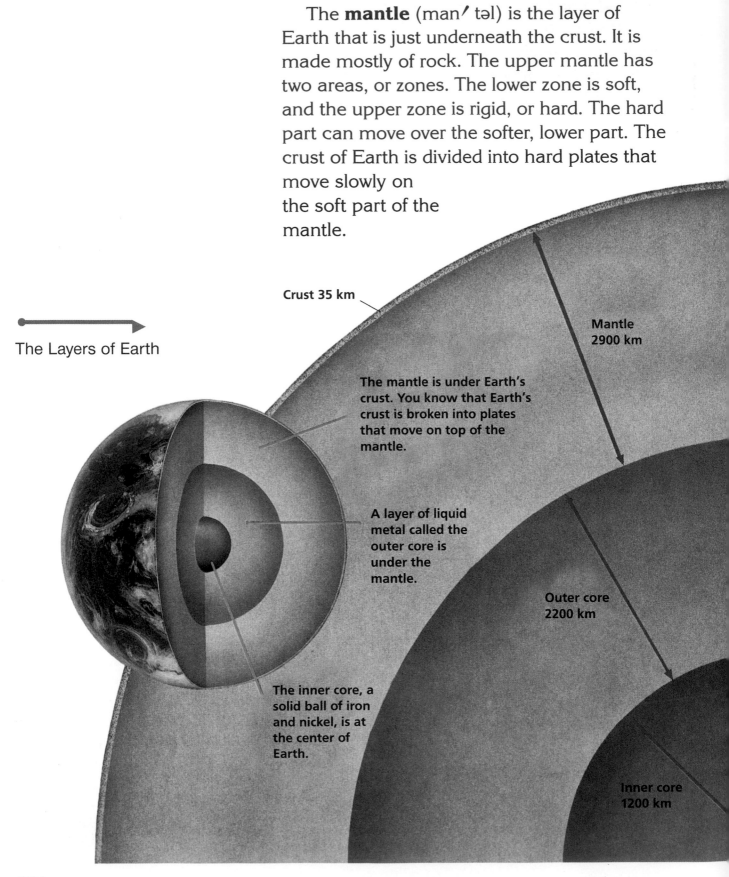

The Layers of Earth

Crust 35 km

Mantle 2900 km

The mantle is under Earth's crust. You know that Earth's crust is broken into plates that move on top of the mantle.

A layer of liquid metal called the outer core is under the mantle.

Outer core 2200 km

The inner core, a solid ball of iron and nickel, is at the center of Earth.

Inner core 1200 km

The center of Earth is called the **core.** It is made up of two layers. The outer core is made of nickel and iron and is so hot that it is molten, or liquid. The inner core is also made of nickel and iron and is even hotter than the outer core. Because of the extreme pressure at Earth's center, the nickel and iron of the inner core become solid.

Lighter materials rose to form the outer core and mantle.

Even lighter materials formed the thin outer crust.

The lightest materials formed an outer layer of gases.

Scientists theorize that Earth formed from a spinning cloud of dust and gases in space. Gravity pulled the more dense iron and nickel inward to form the inner core.

Even though we think rocks on the surface of Earth are heavy, the materials in the inner core are much more dense.

Studying Earth

Scientists have been studying Earth for hundreds of years. A puzzling thing about Earth is that it is always changing. Most of what we can see on the surface looks like it has always been there. But when we study rocks and minerals we find that some rocks are much older than others. This is one way we know that Earth is always changing.

Scientists try to learn how Earth was formed. They study Earth because they want to understand how it changes over time. Scientists have drilled deep into Earth. The deepest they have been able to drill so far is about 12 km.

Inside a cave at Carlsbad Caverns. Underground caves are formed when water, moving through cracks in Earth's crust, dissolves huge areas of underground limestone.

No one has ever been able to dig deep enough to reach Earth's mantle. The outer and inner cores are even deeper. How can scientists know what is deep inside Earth?

They can tell by studying all kinds of waves. For example, ocean waves act a certain way when they hit the shore. They act another way when they are free to travel across the sea. Waves that move through water are just one type of wave.

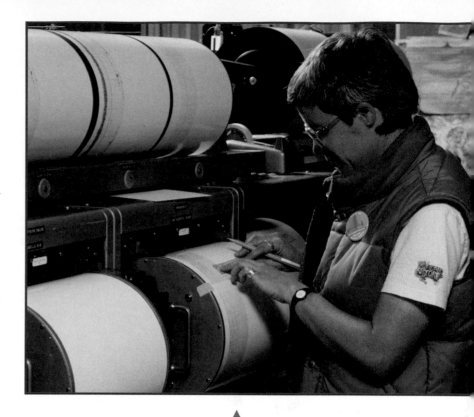

Earthquakes make waves deep inside the ground, too. When scientists study the waves underneath Earth's surface, they notice that earthquake waves act differently depending on what type of material they are traveling through. This helps scientists to know whether an earthquake wave is moving through a solid or a liquid.

Scientists have learned what is inside Earth by studying different kinds of waves.

CHECKPOINT

1. What are Earth's layers called? What are they made of?

2. Describe two ways that scientists study Earth's layers.

 What is inside Earth?

ACTIVITY

Making a Model of Earth's Layers

Find Out

Do this activity to make a model of the layers of Earth.

Process Skills

Constructing Models
Communicating

WHAT YOU NEED

two plastic bags

four shelled round nuts

peanut butter or cream cheese

small ripe banana

two graham crackers

spoon

freezer

Activity Journal

WHAT TO DO

1. Wash your hands. Scoop a spoonful of peanut butter or cream cheese.

2. Place a nut in the middle of the peanut butter or cream cheese. Cover the nut with the peanut butter or cream cheese. Make the peanut butter or cream cheese as thick as the width of the nut.

3. Peel the banana and put it inside the plastic bag. Mash the banana until it is smooth. Place the peanut butter- or cream cheese-covered nut inside a plastic bag. Cover it with a layer of mashed banana that is just a bit thicker than the peanut butter or cream cheese.

4. Put graham crackers into a different plastic bag. Seal the bag. Break the graham crackers into small pieces.

5. Carefully roll the nut that is covered with banana and peanut butter or cream cheese into the mashed graham crackers. Be sure to coat the nut so that the graham crackers cover the banana layer completely.

6. Freeze the whole ball covered in graham crackers for a few hours. The ball is your **model** of Earth.

7. Take one slice out of the nut model. Look at the inside layers very carefully. **Draw** a picture of the layers you see.

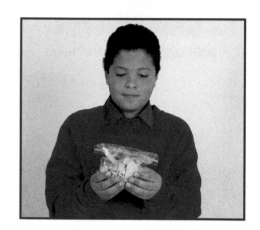

CONCLUSIONS

1. Which layer did the graham cracker represent?

2. Which layer did the banana represent?

3. Which layer did the peanut butter or cream cheese and the nut represent?

ASKING NEW QUESTIONS

1. Where do you think materials in the layers of Earth come from?

SCIENTIFIC METHODS SELF CHECK

✔ Did I **make a model** of Earth's layers?

✔ Did I **communicate** by drawing a picture of my model?

Earth's Forces

Find Out

- What different forces change Earth's surface
- How some landforms change very slowly
- How some landforms change very quickly

Vocabulary

landform
deposition
glacier
volcano
earthquake
faults

The Big QUESTION

What can cause landforms to change?

All kinds of interesting features cover the surface of Earth. Jagged mountain peaks, rounded foothills, and soft, sandy beaches decorate Earth's landscape. How are these features made? Will they always be around?

Land Formation

The shape and features of the land change as time passes. The movements inside Earth, gravity, moving water, and the weather all work together to change the surface features of Earth.

Many changes to the landscape happen over several million years. Some changes are noticed after only a few years, like the formation of a riverbank. Other changes can happen in just a few minutes, like a volcano eruption or a mudslide.

Many forces formed the surface features of Earth.

A feature of Earth's surface, such as a hill, or an island, is called a **landform.** Mountains, plains, valleys, islands, and all the other kinds of land on Earth are landforms.

Landforms are created in many ways. The movement of rocks is one of those ways. Even tiny rock particles can change Earth's surface. When running water, ice, or wind move and deposit small rock particles, this is called **deposition.** The word *deposition* means "to deposit" or "to leave something in a place."

Weathering and climate changes have caused valleys, lakes, and plains. Some of these changes happen quickly. Other changes are slower and harder to see.

Slow Changes

Usually, we do not notice the changes in Earth's landforms because they take place over a long time. The Appalachian Mountains are very old. They began to form more than 400 million years ago! When change happens over that much time, the change happens too slowly to be noticed by people.

Glaciers

A **glacier** (glā′ shər) is a mass of ice that flows slowly over land. A glacier can move about 30 cm a day. When a glacier moves, it carries rocks with it. The pressure of the glacier and the rocks scraping against the land makes new landforms.

Some glaciers were so huge that they covered entire continents, cutting out valleys and other landforms. Almost two million years ago, glaciers covered much of Europe and North America. Glaciers formed the Great Lakes of North America. When the glaciers melted about 12,000 years ago, the water was trapped in the valleys and the Great Lakes were born.

As glaciers move forward, they carry rock fragments underneath them. The ice from a glacier and the rocks dragged with it create new landforms.

Fast Changes

Some landscapes change quickly. Waves crash on beaches, moving sand out to sea. Ocean currents then deposit the sand somewhere else. Heavy storms and floods can move rocks and soil, changing the shape of the land. Rockslides can send rocks down a mountain or a cliff in just seconds. This happens when the rocks are disturbed by weather, vibrations, or other forces. Mudslides can move large amounts of wet mud down a mountain. Earthquakes and volcanoes can change Earth's surface in just a few minutes.

Volcanoes

A **volcano** is a landform created by the eruption of liquid rock from under Earth's crust. Where Earth's crust has a weak place, molten rock from the mantle may force its way to the surface and erupt as lava. Some lava flows almost like water and can cover large areas before it cools and hardens. Other lava is thicker. It usually erupts in explosions that can build tall mountains layer by layer.

A volcano erupting

Even as a mountain is first forming, the wind and weather begin to affect it. As hundreds or thousands of years pass, the surface wears away, slowly changing the mountain's appearance. Some mountains get lower and other mountains become taller as time goes by. One example is the Himalayan mountain range in central Asia. Earth's forces push the mountains up about 2 cm every year.

The San Andreas fault

Earthquake damage in
San Francisco

Earthquakes

An **earthquake** is the sudden movement of Earth's crust. These movements happen when forces within Earth push or pull on rocks. If the forces are strong enough, rocks may suddenly slip along cracks in Earth's crust. These cracks are called **faults.**

Scientists can't predict the exact time, place, or strength of an earthquake. But certain warning signs occur shortly before a large quake strikes. These warning signs can be detected with the use of special instruments.

There are instruments that pick up movement beneath Earth's surface. One instrument uses a laser beam to tell if the land on either side of a fault is moving.

Earthquakes can be measured by scientists. One useful instrument is a *seismograph* (sīz′ mə graf′). It measures earthquake waves, or seismic waves, that move through rocks beneath Earth's surface. Seismic waves cause the rocks of Earth's crust to vibrate. The size or "strength" of an earthquake is measured by using the Richter scale. The Richter scale uses a scale of numbers from one to ten, with one being a very mild earthquake and ten being the most destructive earthquake possible. An earthquake with a measurement greater than five on the Richter scale is strong enough to damage buildings.

A rockslide happens when big and small rocks roll down a mountainside or cliff in just a few seconds.

A mudslide happens when too much rain soaks a mountainside. Gravity pulls wet soil and rocks down.

CHECKPOINT

1. What forces change the shape of land?

2. How can glaciers change Earth's surface?

3. What are volcanoes and earthquakes and how do they change Earth's surface?

 What can cause landforms to change?

ACTIVITY
Moving Ice

Find Out

Do this activity to model how a glacier affects Earth's surface.

Process Skills

Constructing Models
Observing
Measuring
Communicating

WHAT YOU NEED

marker

0.45-kg box of cornstarch

graduated cylinder

gravel

sand

water

waxed paper

large bowl

spoon

Activity Journal

soil

ruler

WHAT TO DO

1. **Make a model** of a glacier. Pour the cornstarch into a large bowl. Add 340 mL of water. Mix until the mixture is thick.

2. Place a sheet of waxed paper on a table. Place a golf ball-sized spoonful of the cornstarch mixture in the middle of the waxed paper. **Observe** the mixture carefully.

3. Place another spoonful of the cornstarch mixture in the center of the model glacier to represent new snow on the glacier. **Observe** how the model reacts.

4. Sprinkle sand, gravel, and soil in a 3-cm band around the edges of the glacier

model. Outline the edge of the band on the waxed paper. Sprinkle a little soil on top of the glacier model.

5. Place a spoonful of the cornstarch mixture on top of the center of the glacier. **Measure** how far the ice cap moves. Repeat until the glacier is 3 cm from the paper's edge.

6. **Draw** your glacier. Compare its thickness at its edges and at its center.

7. Place a second piece of waxed paper over the top of the glacier model. Carefully turn the whole model over so you can see its bottom. **Observe** the position of the sand and soil particles. **Draw** a diagram of your observations.

CONCLUSIONS

1. What happened to the glacier model as it flowed over the sand and soil?

ASKING NEW QUESTIONS

1. How could you set up the model to see how a glacier might behave when it reaches a mountain?

SCIENTIFIC METHODS SELF CHECK

✔ Did I **make a model** glacier?

✔ Did I carefully **observe** the movement of my model glacier and the effects of adding materials to the model?

✔ Did I **measure** the movement of the "ice cap" on my model?

Surface Features of Earth

The Big QUESTION

How are Earth's surface features formed?

What do you see when you look outside? You might be looking at a mountain, rolling hills, or a valley. You might see a flat field in the plains or a desert. You might be looking at the shores of a lake, a river, or an ocean. Wherever you are, you are looking at some type of landform.

Types of Landforms

Planet Earth has many different types of landforms. There are hot, sandy deserts and cold, frozen deserts. There are dense forests, huge mountains, and plains that look like oceans of golden grass. A high plateau often surrounds a wide, deep canyon. There are islands and river deltas and more landforms with unusual names.

Landforms come in three general types. They are mountains, plains, and plateaus. These kinds of landforms make up about one fourth of Earth. The rest of Earth is covered by water.

Mountains

Any landform that is at least 610 m above the ground around it is called a **mountain.** Great forces inside Earth make mountains.

Mountains are formed in many ways. Some mountains are formed when huge pieces of Earth's crust and upper mantle crash into each other. Forces inside Earth cause layers of rock to push up out of the ground. These huge rock layers crumble and fold, forming mountain ridges such as the Appalachian Mountains.

Depending on where you live, you see different landforms. Some places have high mountains and deep valleys. Other places have flat plains or dry deserts. Some places are tropical islands and others are frozen islands. Rivers and lakes are found in many areas. What forces create all these landforms?

B49

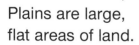

Breaks in Earth's crust called faults helped form the Grand Teton Mountains in the state of Wyoming.

Other mountains are formed when rocks in Earth's crust slide past each other along fault lines. When this happens, large blocks of rock are pushed sideways, up, and down along a fault. This movement forms huge mountains like the Grand Teton Mountains in Wyoming. These are just two of the many ways in which mountains can be formed.

Plains and Plateaus

A **plain** is a wide, mostly flat area of land. Underneath a plain are flat, layered rocks. Plains are formed in different ways.

Glaciers, volcanoes, and water form plains. Glacial plains form when a glacier erodes and leaves soil and rock material behind. Volcanoes form lava plains. When rivers flood their banks and carry soil into valleys, river plains form. Coastal plains form when the ocean retreats. When a lake dries up and its bed is exposed, a lake plain is formed.

Plains are large, flat areas of land.

A plateau is high like a mountain, but its top is wide and flat like a plain.

A **plateau** looks like a combination of a mountain and a plain. It can be as high as a mountain, but its top is flat like a plain. Another name for a plateau is *tableland* because the land is high and flat on the top like a table. Canyons and gorges may form near plateaus. Plateaus form when flat rocks are pushed up by forces under the ground. The rocks do not fold but stay flat.

The Grand Canyon

Sometimes a river will cut through a plateau, creating a canyon. One of the most spectacular examples of this is the Grand Canyon in the southwestern United States. Over millions of years, the Colorado River has cut a path through the Colorado Plateau. The result is the Grand Canyon.

Imagine visiting a new city like San Francisco, California, without knowing your way around town. A city map would help you find your way so you wouldn't get lost.

Topographic Maps

Maps are like models of a place on Earth. They help us measure distances on Earth. They help us find our way. Maps can show the height of mountains and the depths of valleys. Learning to read maps now will help you often in the future.

There are many different kinds of maps. A world map shows us all the countries, islands, oceans, and features of Earth on a small scale. A country map lets us look more closely at one particular country. Some maps show only one town or city. There are maps that show even smaller areas, like a park or the inside of a building.

Maps that show elevation (depth, breadth, and height) are called **topographic maps.** Like other types of maps, topographic maps are models of features on Earth's surface. They use different symbols to show oceans, lakes, rivers, plateaus, mountains, and plains. Topographic maps also show something else. They show height.

A *contour line* is a line that connects points of equal height, or elevation. On a topographic map, elevations are heights above or below sea level. Sea level is exactly what you would guess—it is the average elevation, or level, of the water on an ocean or sea. Sea level is also known as zero elevation. Each contour line on a topographic map shows a different elevation.

A Topographic Map Showing Contour Lines

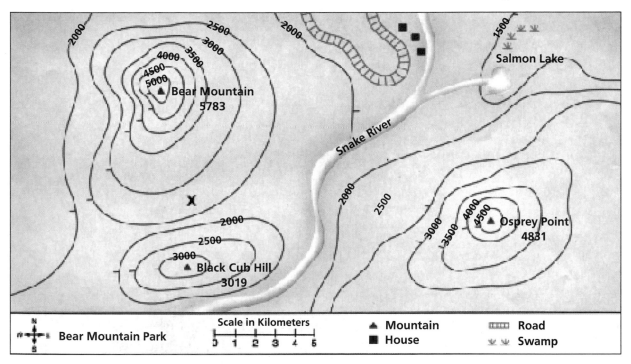

CHECKPOINT

1. Describe some examples of landforms.
2. What is a topographic map?

? How are some of Earth's surface features formed?

ACTIVITY

Modeling How Maps Show Elevation

Find Out

Do this activity to see how the elevation of landforms is shown on a flat piece of paper.

Process Skills

Constructing Models
Measuring
Communicating
Using Numbers

WHAT YOU NEED

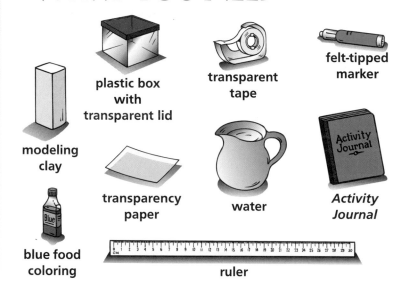

modeling clay

plastic box with transparent lid

transparent tape

felt-tipped marker

transparency paper

water

Activity Journal

blue food coloring

ruler

WHAT TO DO

1. Using modeling clay, **make a model** of a mountain inside the plastic box. Make the mountain's base 10 cm × 10 cm. Make sure the height of the mountain is a little lower than the top of the box. When you are done, **measure** the base and the top of your model. **Record** the measurements.

2. Use the ruler to make a scale in centimeters on the outside of the plastic box. Put the clear lid on the box.

3. Tape the transparency in place on the clear lid. Do not tape the lid closed.

4. Put the box on a flat surface. Fill the box with water to the depth of 1 cm.

5. Look down through the transparency on the clear lid. On the transparency paper, trace the outline of the water's edge as it comes into contact with the mountain model. If it is too difficult to see the water outline, add a few drops of blue food coloring to the water.

6. Label the outline "1 cm."

7. Repeat Steps 4, 5, and 6 by adding water to the depth of 2 cm. Trace the outline on your transparency and label it "2 cm." Do this again with 3 cm, 4 cm, and so on until the water level is even with the top of the mountain model. Make the top line a dotted line on the transparency.

CONCLUSIONS

1. Describe the map you made.

ASKING NEW QUESTIONS

1. How did you show elevation on your map?

2. How can you use a topographic map to determine the height of the mountain?

SCIENTIFIC METHODS SELF CHECK

✔ Did I **make a model** of a mountain and then **measure** the height of my model?

✔ Did I **record** my observations?

Review

Reviewing Vocabulary and Concepts

Write the letter of the word or phrase that completes each sentence.

1. The layer just beneath Earth's crust is called the ___.
   a. core   **b.** dirt
   c. glacier   **d.** mantle

2. Plains, plateaus, and mountains are features of Earth's surface called ___.
   a. deposition   **b.** landforms
   c. volcanoes   **d.** earthquakes

3. When wind, water, or glaciers move and deposit small pieces of rock, it is called ___.
   a. deposition   **b.** tension
   c. eruption   **d.** vibration

4. A mass of ice that flows slowly over land is called ___.
   a. a volcano   **b.** a fault
   c. an earthquake **d.** a glacier

5. Maps that show elevation are called ___.
   a. Richter scales **b.** topographic maps
   c. seismographs   **d.** plates

Match the definition on the left with the correct term.

6. the outer layer of Earth   **a.** plateau

7. center of Earth   **b.** core

8. sudden movement of Earth's crust   **c.** crust

9. wide, mostly flat area of land   **d.** plain

10. land that is high and flat on top **e.** earthquake

Understanding What You Learned

1. Describe how Earth's crust is changing all the time.

2. Describe the layers of Earth from the crust to the core.

3. How did the Great Lakes in North America form?

4. How do plains form?

5. What does zero elevation mean on a topographic map?

Applying What You Learned

1. If you were using a peach as a model of Earth, what would the skin represent?

2. What might you find in an area surrounding a volcano after an eruption?

3. Describe what might happen to a city struck by an earthquake measuring seven on the Richter scale.

4. On a topographic map, what type of landform would be labeled 5500 m?

 5. Do you think the surface of Earth will look the same 5000 years from now as it does today?

For Your *Portfolio*

Think about all you have learned about Earth. Imagine you are a scientist. You have traveled from the top of the mountains to the center of Earth. Write a short story about your journey.

Earth Beneath You

Earth beneath your feet is made up of more things than you may think. There are rocks of different sizes, shapes, and colors. There are different kinds of soils. There are also natural resources under your feet. Some, like coal, we dig up and burn. Others, like silver, we dig up and use to make things.

It may seem that Earth beneath your feet is always the same, but it is not. It is always changing. Rocks are breaking up and forming new rocks. Soil is being created. Resources are being formed, used, and replaced. People depend on everything in Earth beneath our feet.

The Big IDEA

Rocks, soil, and natural resources are found in Earth's surface layers.

CHAPTER SCIENCE INVESTIGATION

Learn about the similarities and differences of soils found in different areas. Find out how in your *Activity Journal.*

B59

Rocks

Find Out

- What rocks are made of
- How the three types of rock form
- What the rock cycle is

Vocabulary

minerals
igneous
sedimentary
metamorphic
rock cycle

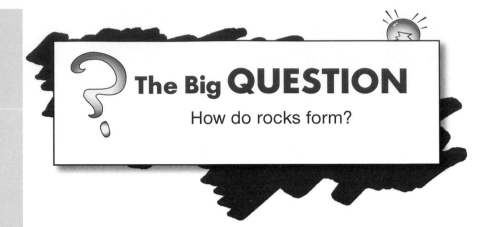

The Big QUESTION

How do rocks form?

Did you know that rocks are around you and beneath you wherever you are? They are in streams, oceans, soil, cities, and parks. They are in mountains, valleys, and fields. They are deep in Earth and on Earth's surface. Have you ever thought about what rocks are made of?

Rocks and Minerals

All rocks are made of one or more minerals. **Minerals** are made of solid, nonliving materials from Earth called elements. Minerals can be made of a single element or compounds of several different elements. Some minerals you might know are silver, diamond, and quartz. The minerals in rocks may be hard or soft, shiny or dull. They may be any color. If you look closely at a rock, you might see more than one color. Each color may be a different mineral.

How Rocks Form

Although you can't always tell, rocks are forming all the time. Rocks are made, broken up, and remade all over Earth. However, not all rocks are formed in the same way. Rocks are classified into three groups based upon how they are formed.

Igneous Rocks

Deep inside Earth, it is so hot that minerals melt. Some of these melted minerals move onto Earth's surface in the form of lava. Lava is hot, molten rock that erupts out of a volcano. It cools on Earth's surface and hardens into rocks. Rocks that form when melted minerals cool and harden are called **igneous** (ig′ nē əs) rocks. There are many different kinds of igneous rocks, both above and below Earth's surface. Some are different because they have different amounts or kinds of minerals in them. Others are different because of how quickly or slowly the rock cools. An example of igneous rock is granite (gran′ it). Granite is hard. It cools slowly below Earth's surface. It is often dug out of the ground and used to make buildings.

One place you can see igneous rocks being made is near a volcano. The lava is melted rock material that comes from deep inside Earth. As it cools on Earth's surface, it forms rocks.

The White Cliffs in Dover, England, are made of a kind of sedimentary rock called chalk. Chalk is made of the mineral calcite and of tiny animals that once lived in the ocean.

Sedimentary Rocks

Another kind of rock forms most often in water. It is called **sedimentary** (sed′ ə men′ tə rē) rock. Do you remember how rocks break up and change because of weathering and erosion? Some bits of rock and minerals, and the remains of once-living plants and animals wash into lakes and oceans.

These bits of material sink, forming layer on top of layer at the bottom of the water. The pressure of the water and the top layers presses down on the layers underneath. The bottom layers are under so much pressure that they slowly harden into sedimentary rock.

There are many different kinds of sedimentary rocks. Two kinds that people use to make buildings are limestone and sandstone.

Another sedimentary rock, coquina (kō kē′ nə), is made of sand and pieces of shells.

Metamorphic Rocks

Metamorphic (met′ ə mor′ fik) rocks are formed when heat and pressure change existing rocks into different kinds of rocks. The very high heat inside Earth can change an existing rock into a metamorphic rock. The pressure and squeezing inside Earth also can change an existing rock into a metamorphic rock.

Granite

Heat and pressure can change the igneous rock granite into gneiss (nīs), a metamorphic rock.

Gneiss

Marble is an example of a metamorphic rock. Marble starts out as limestone, a sedimentary rock. When limestone is deep inside Earth, heat and pressure change it into marble. Marble is harder than limestone. Artists often carve statues and other objects out of marble.

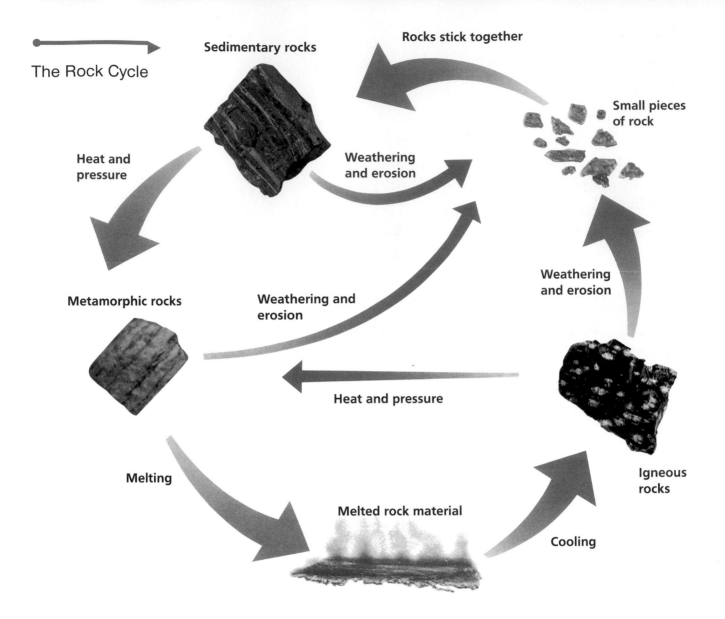

The Rock Cycle

- Sedimentary rocks
- Rocks stick together
- Small pieces of rock
- Heat and pressure
- Weathering and erosion
- Metamorphic rocks
- Weathering and erosion
- Weathering and erosion
- Heat and pressure
- Melting
- Igneous rocks
- Melted rock material
- Cooling

The Rock Cycle

The ground beneath your feet is a busy place. Rocks are constantly forming, changing, and re-forming. Layers of rocks are pushed deep into Earth and are covered by other layers. Some rocks are being squeezed and heated and changed into metamorphic rocks. Melted minerals are rising to the surface of Earth and cooling into igneous rocks. Bits of rocks are being broken off by weathering and erosion. They are washing into lakes, seas, and oceans, where they will slowly become sedimentary rocks. Although most of these things are happening very slowly over millions of years, rocks are always changing.

This constant changing that happens to rocks is called the **rock cycle.** It is called a cycle because it is a series of events that happens over and over without ever stopping. Changes can happen anywhere, in any order, and at any time. Weathering, erosion, melting, cooling, heating, and pressure keep the rock cycle going.

Sand is made of tiny pieces of rocks, coral, and seashells. Sand is an example of the constant changing of rocks in the rock cycle.

CHECKPOINT

1. What are rocks made of?
2. What are the three types of rock?
3. What is the rock cycle?

    ? How do rocks form?

ACTIVITY
Making a Rock Model

WHAT YOU NEED

three colors of modeling clay

waxed paper

plastic knife

Activity Journal

Activity Journal

WHAT TO DO

1. **Make a model** using a number of marble-sized balls from each color of clay.

2. Gently stick different-colored balls together to make a larger ball.

3. On a piece of wax paper, press down on the ball until it is flat. Turn the flattened ball on its edge and press down again.

4. Cut your rock in half with the plastic knife. **Observe** the inside of your rock model.

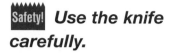 *Use the knife carefully.*

5. Draw what the inside of your rock looks like.

CONCLUSIONS

1. How did the shape of the small, colored balls change?

2. What effect did pressure from your hands have on the shape of the small clay balls?

3. What kind of rock does your model represent?

ASKING NEW QUESTIONS

1. Look at your rock. Would it be easier to separate the colored balls now or before you pressed the rock model together?

2. How might heat change your rock?

SCIENTIFIC METHODS SELF CHECK

✔ Did I **make a model** of the rock?

✔ Did I **observe** what the model looked like?

✔ Did I **record** my observation?

Soils

Find Out

- How soil is formed
- What the three main types of soil are
- Why soil is important

Vocabulary

soil
humus
clay soil
sandy soil
loam

The Big QUESTION

How do soils differ from place to place?

Climate differs from place to place on Earth. The way people live also differs, such as what they grow to eat and how they build their homes. One thing that affects what people grow and sometimes what their homes are made of is soil. Soil is a very important part of Earth beneath your feet.

How Soil Forms

Soil is part of the top layer of Earth's surface. Soil is made of tiny bits of broken rock and decaying plant and animal matter.

Rocks on Earth's surface contribute to soil when they are broken apart by weathering. Wind, rain, ice, and the movement of water can cause a rock to break apart. Over thousands of years, the rock slowly crumbles and becomes a part of soil.

Another part of soil comes from plant and animal material. When plants and animals die, they decay. This decaying material becomes an important part of soil. It is called **humus** (hyōo′ məs). Humus gives soil a dark color. Plants need humus to live and grow.

People, plants, and animals need soil to live.

Plants and animals that are alive also help soil form. The roots of trees can grow into cracks in rocks and force the rocks to break apart. The tunnels that animals make in soil let air and water into it. Air and water help soil form faster. Earthworms are especially good at helping soil form. Beneath a square kilometer of grassland, there may be thousands of earthworms tunneling through the soil.

Soil does not form in an even layer all over Earth. If the land is flat, a deep layer of soil may form. A thinner layer may form on hills and mountains. That is because each time it rains, some soil is washed downhill and eroded away with the water. In some areas, such as very rocky places, there may be no soil at all.

The tunnels that earthworms make help soil form.

How Soil Varies

The soil in one place may be very different from the soil in another place. Soil can be black, red, brown, gray, or yellow. It can have coarse grains or fine grains. It can be smooth or gritty. It all depends on what is in the soil.

Soil can have different kinds of rocks and minerals in it. Some minerals give soil its color. The mineral iron can give soil a red color. Soil can have a lot of humus in it or no humus at all. Soil that contains a lot of humus is dark brown or black.

Soil can also have different amounts of air and water in it. Soil that has a lot of water in it will feel damp when you touch it. Soil with many animal tunnels through it often has more air than soil without animal tunnels.

Soil may contain other things, too. It may contain rocks or pebbles. It may contain decaying pieces of plants and animals that haven't changed into humus yet. It may contain insects and insect eggs. In many places, soil is home to earthworms, moles, insects, and groundhogs.

There are three main types of soil. **Clay soil** has tiny grains in it. It feels smooth. Clay soil holds water well. It can become hard and packed together. Many plants can't grow in clay soil.

Sandy soil has large grains. It feels gritty. Water flows easily through sandy soil. Sandy soil is loose and easy to dig. Only certain plants grow well in sandy soil.

Loam (lōm) is a mixture of clay, sand, and humus. It feels soft. Loam holds water well. Many kinds of plants grow well in loam.

Each of the three types of soil has a different look and feel. When you look at a sample of soil, you will often see two types mixed together. A sample that has a lot of one type of soil is "rich" in that type. For example, soil may be rich in clay.

A Soil Chart

Soil Type	What Grows in It	Color	Texture	Ability to Hold Water	Humus Content
Clay	Lush plant life in jungles	Red or brown	Tightly packed soil, with tiny grains; sticky when moist	Holds water	Some humus
Loam	Good for crops and trees	Brown or black	Mixture of fine and coarse materials	Holds water well	A great amount of humus
Sand	Small trees and cacti in warm dry places	Gray or yellow-white	Loose soil with large grains; gritty	Holds very little water	Little humus

The Importance of Soil

People, plants, and animals need soil to live. Plants need soil to grow, and people need plants for food. Some materials people use, such as cotton for clothes, also come from plants. Trees need soil to grow. People use trees to make paper, to build houses, to make furniture, and even to make some types of fabric for clothing.

Animals also need soil to live. Some animals eat plants that grow in soil. People eat the meat, eggs, and milk from these animals. Other animals live in the soil itself. Soil contributes in many ways to our way of life on Earth.

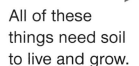

All of these things need soil to live and grow.

Corn grows in loam on a farm in Pennsylvania.

Lemon trees grow in the sandy soil of a California orchard.

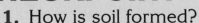

Pecan trees thrive in Georgia clay.

CHECKPOINT

1. How is soil formed?
2. What are the three main types of soil?
3. Why is soil important?
 How do soils differ from place to place?

ACTIVITY
Observing Soil Types

Find Out

Do this activity to learn what some differences are between three types of soil.

Process Skills

Predicting
Observing
Communicating

WHAT YOU NEED

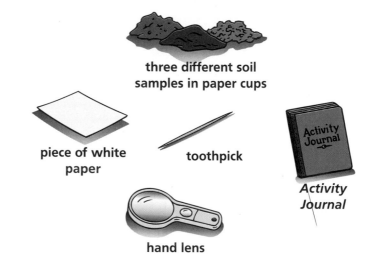

three different soil samples in paper cups

piece of white paper

toothpick

Activity Journal

hand lens

WHAT TO DO

1. Sprinkle a small amount of each soil sample on the paper. **Observe** each soil sample.

2. **Predict** what you will see in each sample when you use a hand lens.

3. Use the toothpick to separate bits of soil into piles according to what they look like.

4. **Observe** each soil sample with a hand lens. Notice the color, size, and shape of the bits.

Safety! *Keep your hands away from your face, and wash your hands after handling the soil.*

5. Feel the soil. Feel to see if it is coarse, gritty, smooth, hard, soft, wet, or dry.

6. **Record** what you observe.

CONCLUSIONS

1. What different kinds of materials did you find?

2. Are any of your samples sticky like clay or gritty like sand?

ASKING NEW QUESTIONS

1. Which of your soil samples do you think would be best for growing plants?

2. Which of your soil samples do you think would hold the most water? How can you find out?

SCIENTIFIC METHODS SELF CHECK

✔ Did I **predict** what I would see?

✔ Did I **observe** how the soil looked?

✔ Did I **record** my observations?

Natural Resources

Find Out

- What some different kinds of natural resources are
- Why natural resources need to be protected

Vocabulary

resource
renewable
nonrenewable
inexhaustible

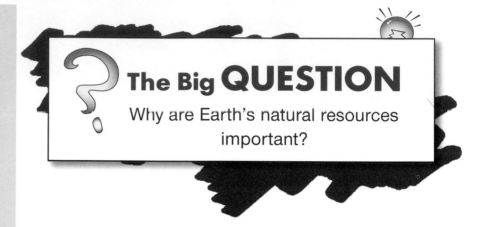

The Big QUESTION

Why are Earth's natural resources important?

You, like all living things, depend on Earth to provide what you need to live. The food you eat comes from the sun's energy, plants, and animals. There is plenty of water for you to drink. There is oxygen in the air for you to breathe. How can we protect these precious things that give us life?

All living things depend on Earth's natural resources.

Earth's Natural Resources

A natural **resource** is a material found in nature that is useful or necessary to living things. Water, oxygen, soil, trees, and minerals in rocks are natural resources. Why do you think we should be concerned about protecting natural resources?

Renewable Resources

Some natural resources are renewable. A **renewable** resource is something that can be replaced within a span of time that is useful to people—about 30 years. Water, air, trees, fish, and animals are renewable resources. People need and use a lot of renewable resources all over Earth. For example, freshwater is necessary for people to live, and people use wood from trees for many products.

Sometimes what people do harms a renewable resource. We use the resource faster than it can replace itself. This has happened to fish in some places. People caught too many fish for food. There weren't enough fish left to reproduce to replace the fish that were caught.

Fish are an example of a renewable resource. People need to be careful not to use too much of natural resources such as fish.

Nonrenewable Resources

A **nonrenewable** resource is something that can't be replaced in a span of time that is useful for people. Soil, rocks, and minerals are examples of nonrenewable resources. They take a very long time (thousands of years or more) to develop. Their supply is limited. They can be used up.

People use many different kinds of nonrenewable resources. One kind is coal, a sedimentary rock. Coal is an important resource that is used for heat and power. It is mined all over the world. Another kind is petroleum (pə trōl′ ē əm), a mixture of gases, liquids, and solids found under Earth's surface. It can be separated into several useful products, including natural gas, gasoline, and fuel oil.

Minerals like copper, lead, and silver are nonrenewable resources. Copper is used to make electric wires and coins. Lead is used to make batteries. Silver is made into mirrors, film, and jewelry. To get to minerals, miners dig deep tunnels or they remove the soil to form large open pits. Nothing will grow in the open pits unless the soil and some minerals are replaced.

An open-pit copper mine in Utah

Inexhaustible Resources

A few of Earth's natural resources are called **inexhaustible** (in eg zos′ tə bəl) resources. They can never be used up. The sun and wind are both inexhaustible resources. Others are the heat deep in Earth and the natural power of running water.

People all over Earth have been using inexhaustible resources for hundreds of years. Windmills pump water and generate electric power. Flowing water can turn waterwheels to grind grain.

Recently, people have begun to use inexhaustible resources in new ways. In sunny places, solar panels collect solar energy that is converted to heat and power. In windy places, wind generators convert wind energy to electricity for entire towns. Using inexhaustible resources for energy means we won't use so many nonrenewable resources, which will last longer.

One large wind generator like the ones shown here can produce enough electricity for over 250 homes.

Protecting Earth's Resources

Replanting a forest

When you look around outside, Earth seems huge. It seems like there should be enough resources for everyone. But there are many people now using Earth's resources. People can cause the air and the water to become polluted. People are using up some resources too quickly and making others unusable. For living things to survive, we must protect Earth's natural resources from being used up or destroyed.

People are working in many ways to protect Earth's resources. Some are trying to figure out how to reduce water pollution. Others help save our freshwater resource by using less of it. People are also trying to protect the air by reducing the pollution from cars and factories. Others help protect the air by replanting forests. These people know that trees are important for cleaning the air and releasing oxygen into the air.

You can do many things to help protect Earth's resources. To save water, you can turn off the faucet when you brush your teeth. You can recycle and find other uses for paper, cans, plastics, and glass. Because we all depend on Earth, everyone needs to help protect Earth's resources.

The United Nations is an international organization that has made protecting natural resources one of its main concerns.

CHECKPOINT

1. What different kinds of natural resources are there?

2. Why do people need to protect Earth's resources?

 Why are Earth's natural resources important?

ACTIVITY
Classifying Resources

Find Out

Do this activity to see what resources you and your family use.

Process Skills

Predicting
Observing
Classifying
Communicating

WHAT YOU NEED

gloves

two grocery bags

one day's worth
of trash

pencil

Activity
Journal

WHAT TO DO

1. For one day, collect all the trash thrown away in your home or classroom. **Predict** what most of the trash will be made of.

   Safety! *Wear gloves when handling the trash. Make sure there is no broken glass in the trash before you handle it.*

2. Put everything made of glass, aluminum, other metals, or plastic in one grocery bag. Put everything else in the other bag.

3. **Observe** the items that you collected. Was your prediction correct? List the contents of each bag, grouping them with other items made of the same materials.

4. Next to each item on your list, **write** the name of the resource that was used to make that item. Then, **classify** the resource as renewable or nonrenewable.

CONCLUSIONS

1. Which things on your list can be recycled?

2. What might be made from the recycled items?

ASKING NEW QUESTIONS

1. What changes in your habits could you make to help protect natural resources?

2. What suggestions could you make to others to promote more recycling?

SCIENTIFIC METHODS SELF CHECK

✓ Did I **predict** what I would find?

✓ Did I **record** what I **observed?**

✓ Did I **classify** the materials I collected by writing both the materials they are made of and the type of resource used to make them?

Review

Reviewing Vocabulary and Concepts

Write the letter of the word or phrase that completes each sentence.

1. Rocks formed when melted minerals cool are ___.
   a. igneous **b.** loam
   c. metamorphic **d.** coquina

2. The kind of rock that often forms in water is ___.
   a. igneous **b.** sedimentary
   c. metamorphic **d.** mineral

3. Rocks formed when heat and pressure change existing rocks are called ___.
   a. chalk **b.** sedimentary
   c. metamorphic **d.** clay

4. A material found in nature that is useful to living things is called___.
   a. humus **b.** a natural resource
   c. a rock cycle **d.** loam

5. A material that can never be used up is ___.
   a. a renewable resource **b.** soil
   c. an inexhaustible resource **d.** a nonrenewable resource

Match the definition on the left with the correct term.

6. substances made of nonliving materials from Earth **a.** minerals

7. constant changing of rocks **b.** loam

8. part of the top layer of Earth **c.** rock cycle

9. decaying material that becomes part of the soil **d.** soil

10. mixture of clay, sand, and humus **e.** humus

Understanding What You Learned

1. What is needed for metamorphic rocks to form?

2. What things keep the rock cycle going?

3. What is soil made of?

4. List some renewable resources.

5. What is a nonrenewable resource?

Applying What You Learned

1. Compare and contrast granite, limestone, and marble.

2. What kind of soil would you use if you were going to plant a garden?

3. Is a renewable resource inexhaustible? Why or why not?

4. How are people using natural resources wisely when they take newspapers to be recycled?

 5. Name three important things that can be found in Earth's surface.

For Your **Portfolio**

Think about a way in which people can stop wasting minerals or soil. Make a poster that shows your idea. At the bottom of your poster, write a short paragraph describing your idea.

Unit Review

Concept Review

1. What do we know about the makeup of our solar system?

2. Describe how Earth's crust and landforms have changed over a very long period of time.

3. Explain why rocks and soil are natural resources.

Problem Solving

1. What project might you create to help a younger child understand many of the things you have learned about our solar system?

2. If you were a rock at the top of a mountain, how might you change and move over a very long period of time?

3. Which would hold water better—a beach or a planted field?

Something to Do

With a group, make three mobiles to hang in the classroom. Use paper and other materials in the classroom to create your mobiles. Make one mobile of the solar system. Label the planets, the moon, and the sun. You might include some other stars and constellations too. Create a second mobile of Earth with a section cut out so that you can see the different layers. Show the continents and landforms by labeling and coloring them. For your third mobile, show the different types of rocks. Label the rocks and color them correctly.

UNIT C

Physical Science

1 How Matter Changes

Everywhere you look, you can see matter. Matter includes solid things like rocks, liquids like milk, and gases like the air you breathe. You and your friends are made of matter, too.

People are still learning about matter. Because of what we have learned, we have been able to cure deadly diseases and send shuttles into outer space. Learning about matter will help you to understand the changes you see in the world around you. In this chapter you will learn how to describe and change matter. You will discover what can happen when you combine different kinds of matter.

The Big IDEA

Changes can occur when you combine different kinds of matter.

CHAPTER SCIENCE INVESTIGATION

You can separate solutions and mixtures. Find out how in your *Activity Journal.*

Properties of Matter

The Big QUESTION

How can you describe matter?

*R*ain, snow, the wind in your hair.

Pencils, computers, even your chair.

Things you can feel and things you can see.

What are things made of? What could it be?

Tiny Pieces of Matter

Look at the picture of gold bars. Gold is made up of only one kind of matter: gold. Imagine that you could cut one of these gold bars with scissors into smaller and smaller pieces. If you could cut and cut down to the smallest piece you could see with a hand lens, what would you have? Even the tiniest speck of gold is made up of millions and millions of smaller pieces called atoms.

An **atom** (at′ əm) is a very tiny piece of any kind of matter. In fact, atoms are so small that you cannot see them with your eyes. Even with the most powerful microscopes on Earth, it is difficult for scientists to clearly see an atom.

Scientists sometimes refer to atoms as the building blocks of matter. If you have ever used wooden blocks to build a wall or a building, you can guess what the scientists mean. Your wall was made of many wooden blocks. In the same way, each of the gold bars is made up of billions and billions of atoms stacked together.

Gold is a very valuable metal. It is used in making jewelry and computer chips.

How Elements Form

Scientists know that different kinds of atoms exist. Billions of iron atoms stack together to form an iron nail. Billions of copper atoms stack together to form a copper wire. Matter, such as iron or copper, that is made up of only one kind of atom is called an **element** (el′ ə mənt).

Scientists have identified over a hundred elements. Some elements, such as silver and copper, are metals. Other elements, such as **oxygen** (ok′ si jən), are gases found in the air you breathe. All the different elements have one thing in common. Each is made of only one kind of atom.

People once thought that earth, wind, fire, and water were the basic elements that made up all matter. Now we know that there are over 100 different types of atoms, which are displayed on the Periodic Table of the Elements shown here. Each element has an abbreviation.

Periodic Table of the Elements

	1	2	3	4	5	6	7	8	9	10	11	12	13	14	15	16	17	18
1	hydrogen **H**																	helium **He**
2	lithium **Li**	beryllium **Be**											boron **B**	carbon **C**	nitrogen **N**	oxygen **O**	fluorine **F**	neon **Ne**
3	sodium **Na**	magnesium **Mg**											aluminum **Al**	silicon **Si**	phosphorus **P**	sulfur **S**	chlorine **Cl**	argon **Ar**
4	potassium **K**	calcium **Ca**	scandium **Sc**	titanium **Ti**	vanadium **V**	chromium **Cr**	manganese **Mn**	iron **Fe**	cobalt **Co**	nickel **Ni**	copper **Cu**	zinc **Zn**	gallium **Ga**	germanium **Ge**	arsenic **As**	selenium **Se**	bromine **Br**	krypton **Kr**
5	rubidium **Rb**	strontium **Sr**	yttrium **Y**	zirconium **Zr**	niobium **Nb**	molybdenum **Mo**	technetium **Tc**	ruthenium **Ru**	rhodium **Rh**	palladium **Pd**	silver **Ag**	cadmium **Cd**	indium **In**	tin **Sn**	antimony **Sb**	tellurium **Te**	iodine **I**	xenon **Xe**
6	cesium **Cs**	barium **Ba**	lanthanum **La**	hafnium **Hf**	tantalum **Ta**	tungsten **W**	rhenium **Re**	osmium **Os**	iridium **Ir**	platinum **Pt**	gold **Au**	mercury **Hg**	thallium **Tl**	lead **Pb**	bismuth **Bi**	polonium **Po**	astatine **At**	radon **Rn**
7	francium **Fr**	radium **Ra**	actinium **Ac**	rutherfordium **Rf**	dubnium **Db**	seaborgium **Sg**	bohrium **Bh**	hassium **Hs**	meitnerium **Mt**	(unnamed) **Uun**	(unnamed) **Uuu**	(unnamed) **Uub**						

Key:
element name
element symbol

Lanthanide Series

cerium **Ce**	praseodymium **Pr**	neodymium **Nd**	promethium **Pm**	samarium **Sm**	europium **Eu**	gadolinium **Gd**	terbium **Tb**	dysprosium **Dy**	holmium **Ho**	erbium **Er**	thulium **Tm**	ytterbium **Yb**	lutetium **Lu**

Actinide Series

thorium **Th**	protactinium **Pa**	uranium **U**	neptunium **Np**	plutonium **Pu**	americium **Am**	curium **Cm**	berkelium **Bk**	californium **Cf**	einsteinium **Es**	fermium **Fm**	mendelevium **Md**	nobelium **No**	lawrencium **Lr**

States of Matter

Scientists classify matter according to its state, or form. Three states of matter are solids, liquids, and gases. Each state has certain properties. A **property** (prop′ ur tē) is a characteristic feature. You can describe matter by its properties. For example, a grape is black, purple, red, or green. It is round like a ball. Describing properties is a way of describing matter.

The tennis ball is larger than the ball bearing. If you held both of them, though, the ball bearing would feel heavier.

Properties of Solid Objects

Look at the things in the picture at the top of this page. Each is a solid object. Think about the color, shape, and size of each object. Color, shape, and size are properties of solid objects. Another property of solids is texture, or how the object feels. The texture of the steel ball bearing is smooth. The tennis ball has a rough texture.

Each solid object has a shape and size that doesn't easily change. If you placed the mug in a sink full of water, the size and shape of the mug would not change. If you put the ball bearing in the freezer, it would have the same shape the next day.

Properties of Liquids

Liquids also have certain properties. You can describe a liquid by its color. Honey is golden. Cranberry juice is red. You can also describe the texture of a liquid. Honey is sticky. Water is clear and wet.

Matter can also be a liquid. Here are some liquids that you might see every day.

Gases, like liquids, take the shape of their container.

Liquids are different from solids in an important way. A liquid has no shape of its own. If you poured the water from the glass into a bowl, the water would take the shape of the bowl. The amount of water would not change, but its shape would.

Properties of Gases

A third form of matter is gas. The air you breathe is a mixture of mainly two different elements that are gases, oxygen and nitrogen. Air has no color. You can't smell it. It has no shape of its own. Still, you know that air is all around you. All you have to do is breathe in deeply and feel the air fill your lungs.

There are many different kinds of gases in the world all around you. Have you ever seen a bunch of balloons floating in the air? The gas helium is put inside party balloons to make them float. The bubbles in soda pop and the air in bicycle tires are also made of gases.

Changing States of Matter

Do you know that a solid can change to a liquid? A liquid can also change to a gas. This can happen because the atoms in every kind of matter are always moving. You can't see them moving, but they are. Even the atoms in your chair are moving.

The atoms in solid objects are packed tightly together in an orderly pattern, like bricks in a brick wall. They can just barely move. The atoms in a liquid are not held together in a fixed pattern. They move in a jumbled, disorderly way. That is why a liquid has no shape and why you can pour liquids. The atoms in a gas are very far apart compared to the atoms in solids and liquids. These atoms move around easily and quickly.

For a solid to change to a liquid, the atoms in the solid have to move around more quickly. For a liquid to change to a gas, the atoms have to speed up even more. Heat causes this to happen. Heat causes atoms to move more freely and more quickly.

A Familiar Example

Water is made up of atoms of two different elements—hydrogen and oxygen. Ice is water in its solid state. If you place an ice cube on a plate in the kitchen, the ice will melt and change to liquid water.

If you place the plate of water in the sun for several hours, the water will dry up. The sun warms the water. This causes the bunches of atoms in the liquid to move quickly. When bunches of hydrogen and oxygen atoms are moving very quickly, they break free of one another. The liquid then turns to a gas. The change of a liquid to a gas is called **evaporation** (i vap′ ə rā shən).

Ice isn't the only solid that can turn into a liquid. Solid rock can become so hot inside Earth that the rock turns to liquid. This stream of liquid rock, or lava, is flowing from a volcano.

CHECKPOINT

1. What do all forms of matter have in common?

2. What are three forms of matter?

3. Explain how an ice cube can disappear.

 How can you describe matter?

ACTIVITY

Describing Matter

WHAT YOU NEED

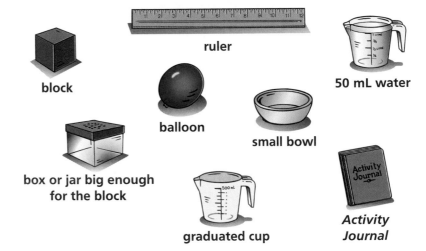

ruler

block

50 mL water

balloon

small bowl

box or jar big enough
for the block

graduated cup

*Activity
Journal*

WHAT TO DO

1. **Observe** the block. What shape is it? **Measure** its length and width. What color is the block? How does the texture of the block feel? Put the block into an empty container. What happens to the shape of the block?

2. Pour 50 mL of water into the measuring cup. What shape does the water have? Now, pour the water into the bowl. What happened to the shape of the water? Pour the water back into the measuring cup, being careful not to spill any of it. Has the amount of water changed?

3. Blow up the balloon halfway. Hold the end so that the air does not escape. How would you describe the shape of the air? Gently squeeze the balloon. Can you squeeze the balloon into a smaller shape? Can you change the shape of the air?

4. **Record** your observations.

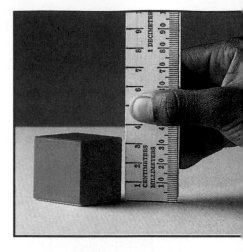

CONCLUSIONS

1. How was the water different from the block?
2. How was the air inside the balloon different from both the water and the block? Was it easy to change the size and shape of the air?
3. Think about the activity you just did. Complete a chart of characteristics of five different objects.

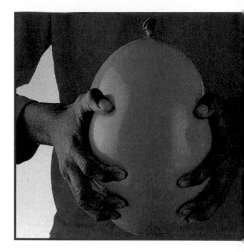

ASKING NEW QUESTIONS

1. What shape does air take when you fill a bicycle tire?
2. Could you pump more air into the bicycle tire without making the tire bigger?
3. List the five objects in the chart you completed. Does each object exist as a solid, a liquid, or a gas? Make a check mark in the columns that apply. Compare your chart with a classmate's chart.

SCIENTIFIC METHODS SELF CHECK

✔ Did I **observe** three different types of matter?

✔ Did I **measure** both the block and the amount of water?

✔ Did I **record** my observations?

Combining Substances

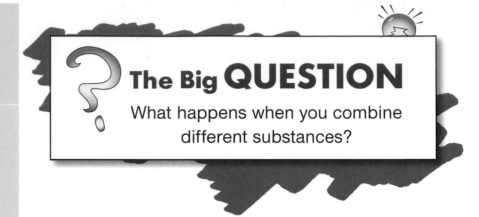

The Big QUESTION

What happens when you combine different substances?

What's your favorite mixture of food? Some people love spaghetti and meatballs. Other favorites include ham and cheese sandwiches and chocolate milkshakes. In this lesson you will learn that foods are not the only things that you can mix together.

Elements and Compounds

Can you name some elements? Silver and oxygen are elements. So are gold and iron. All elements have one thing in common. All the atoms of each element are the same. In silver all the silver atoms are the same. In oxygen all the oxygen atoms are the same.

Elements have another thing in common. Every sample of an element, no matter where it is found, has the same properties. Silver found in the United States is the same as silver found anywhere else in the world.

All pieces will have the same shiny surface and feel the same. The oxygen in the air you breathe is the same as the oxygen in China.

Matter can be made up of either a single kind of atom or many different kinds of atoms joined together. When billions and billions of the same kind of atom are joined together to form matter, that matter is an element. Elements always have the same name as the atoms that form them. For example, the element zinc is made up only of zinc atoms.

Matter can also be a **compound.** A compound (kom′ pound) forms when two or more atoms of different elements combine to make matter. The salt you shake onto food is a compound. Table salt is made of the elements sodium (sō′ dē əm) and chlorine (klōr′ ēn). Water is also a compound. Water is made of the elements oxygen and hydrogen.

Sodium + Chlorine =

A sodium atom and a chlorine atom join together to form table salt.

Like elements, every sample of a compound has the same properties. The table salt you use is the same as the table salt used everywhere else. It has the same taste and looks the same.

Scientists call both elements and compounds substances. **Substances** (sub′ stans iz) are basic materials that make up all things. You can mix different substances together. In this lesson, you are going to find out what can happen when you combine substances.

Combining Substances

A **mixture** (miks′ chər) is a combination of two or more substances. When you wash your hands, you mix soap and water together. When you pour milk on cereal, you make a mixture of milk and cereal. Chocolate milk is a mixture of chocolate syrup and milk.

From these examples, you can see that in a mixture, the same thing doesn't always happen. A bar of soap and water form suds. In a bowl of cereal, the milk and cereal remain separate. Chocolate syrup blends with the milk, and the mixture turns dark. When you mix substances, different things can happen.

In some mixtures, you can see the substances you've combined. Each substance still has the properties it had before you mixed them. Can you guess one property of iron? That's right. Iron is attracted by magnets.

Mixtures

Sometimes when you try to mix two substances together, nothing happens. If you pour a spoonful of cooking oil into a glass of water, the oil will float on the surface. Stirring the mixture with a spoon has no effect. After the water stops swirling, the oil still floats on the surface. Oil and water don't mix.

In a mixture, two or more kinds of matter are mixed together, but they keep their own properties. If you mix iron filings with sand, you will get a rough mixture. Bits of sand will be mixed with the iron filings. You could separate the two substances. Do you know how? If you dipped a magnet into the mixture, the iron filings would be attracted to the magnet. If you keep doing this, you could remove all the iron filings. This mixture is easy to separate because the iron filings and the sand have different properties. One property of the iron filings is that they are attracted by magnets.

Solutions

Now let's look at a different kind of mixture. If you poured a spoonful of table salt into a glass of water and stirred it, what would happen? The salt would dissolve. You could still see a liquid that looks like water, but you couldn't see the salt. This kind of mixture is called a solution. A **solution** (sə lōō′ shən) is a mixture formed when one substance dissolves in another substance.

You could separate the salt from the water. First, you could pour a small amount of the solution on a plate. Then, you'd place the plate in the sun and wait a few hours. The water would evaporate. Salt would be left on the plate.

The change that occurs when something dissolves is called a physical change. In a physical change, no new substance forms. The salt and water solution is still made up only of salt and water. The salt simply dissolved into the water. No new substance was formed in the mixture.

This lemonade was made by mixing water, lemon juice, and sugar. Is it a solution or a mixture? It's a solution because the sugar dissolves in the liquid.

Chemical Changes Produce New Substances

You've probably seen shiny new pennies many times. But most of the pennies you see are not shiny. They are dull. These dull pennies were once shiny, too. What happened to them? A chemical change took place on the surface of the pennies. During a chemical change, new substances are formed. The brown substance that forms on pennies is called tarnish.

You can remove tarnish from a penny in several ways. You can scratch it off. This would take a long time, though. You can also remove the tarnish by causing another chemical change. You could do this by putting the tarnished penny in a small dish of lemon juice. After ten minutes, the tarnish would be gone! The penny would be shiny again. Substances in the lemon juice combine with the tarnish and cause a chemical change.

The brown material on the darker penny is made up of new substances formed when copper combines with other substances from people's hands or the air.

Other Examples of Chemical Changes

You probably see the results of chemical changes every day. Have you ever tasted sour milk? Have you ever seen rust on metal objects? Have you ever seen the dark ash that remains after something burns? These are all results of chemical changes that take place when substances are combined.

Substances combined to cause the rust on this car. Moisture in the air mixed with substances in the metal to create a chemical change.

CHECKPOINT

1. How are elements and compounds different? How are they alike?

2. What is a mixture? A solution?

3. How is a chemical change different from a physical change?

 What happens when you combine different substances?

ACTIVITY

Observing a Chemical Change

Find Out

Do this activity to learn how to combine substances to cause a chemical change.

Process Skills

Predicting
Communicating
Observing

WHAT YOU NEED

three cups

three pieces of steel wool

water

cooking oil

Activity Journal

WHAT TO DO

1. Dip one piece of steel wool in water. Dip the second piece in cooking oil and then in water. Do nothing to the third piece.

2. Put each piece of steel wool in a separate cup. Place the cups where they won't be disturbed.

3. Wait two days. **Observe** any changes in the pieces of steel wool. Was your prediction correct? **Record** your observations.

4. **Predict** what each piece of steel wool will look like after two more days have passed. **Record** your predictions.

5. **Observe** the steel wool pieces again after two more days. **Record** your observations.

CONCLUSIONS

1. What kinds of matter formed the rust?
2. Is rust the result of a physical change or a chemical change? Why?
3. Compare your predictions about changes to the steel wool to your observations.

ASKING NEW QUESTIONS

1. What other objects around you do you think might rust?
2. What would need to happen before rust could form on these objects?
3. How do you think you could prevent rust?

SCIENTIFIC METHODS SELF CHECK

✔ Did I **predict** which of the pieces of steel wool would rust?

✔ Did I **observe** the changes in the steel wool samples?

✔ Did I **record** my observations?

Energy

Find Out

- What stored energy is
- How energy changes form
- How people use energy

Vocabulary

energy
heat
fuels

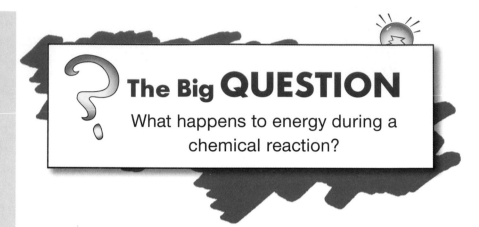

The Big QUESTION

What happens to energy during a chemical reaction?

Every day you use energy to do things. When you get out of bed in the morning, you use energy. When you brush your teeth, you use energy. When you open a book, you use energy. Everything you do involves energy. But what is energy?

Energy is the ability to make things move or change. You know that your body has energy because you can move. You moved your arms and legs when you got out of bed. You can also change things. You changed this book when you opened it.

Stored Energy

But what about when you are not moving or changing anything? What if you are asleep and not moving? Do you still have energy? Yes. Your body is constantly using energy just to stay alive. But there is a lot more energy inside your body than you could ever use at one time. Energy in your body that is not being used is called stored energy. Before a race, runners have a lot of stored energy. When they start to run, they use energy. At the end of the race, they have less stored energy because they used some of it while running.

Other Sources of Energy

You are not the only thing that has energy. In fact, everything has energy. All living things have energy. And all nonliving things have energy, too. What happens when you wind up a toy? The spring in the toy stores energy that will make the toy move when you let it go. What happens when you light a candle? It burns and lights up the space around it. The wick and the wax of the candle have stored energy that is used when the candle burns.

There are many different kinds and forms of matter. There are also different forms of energy. These different forms of energy can change to other forms of energy. In this lesson, you will see how energy can change form.

When you wind up this toy, it has stored energy. When you let the toy go, it moves and uses energy, changing its stored energy into motion.

The stored energy of the jet fuel is released as light and heat when it burns. This energy causes the airplane to move forward at high speed.

How Energy Changes Form

When someone strikes a match, substances on the head of the match combine with substances in the air. This chemical change creates fire. The mixture causes a chemical change. When you light a candle, substances in the candle wax react with the oxygen in the air. The candle burns. It gives off light. Light is a form of energy that lets us see things.

The flame also gives off heat. The heat from the flame makes the nitrogen and oxygen atoms in the air move faster. The faster the atoms move, the higher the temperature. **Heat** is energy moving from a warmer object to a cooler object. Heat always flows from the warmer object to the cooler object—never the other way around.

Substances that we burn in order to release heat are called **fuels.** Other examples of fuels are coal, wood, oil, and gasoline. Coal and fuel oil are burned to heat some homes and buildings. Wood is burned in fireplaces for heat and to cook food. Gasoline is burned in car engines.

The Energy in Batteries

Burning fuel is not the only way to release heat. Batteries can release heat and other kinds of energy. Look at the drawing of the flashlight. Substances in the batteries mix to cause a chemical change. This chemical change releases heat. Chemical changes cause the tiny wire inside the lightbulb to heat up until it begins to glow. The

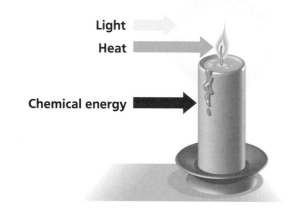

Heat and light are forms of energy.

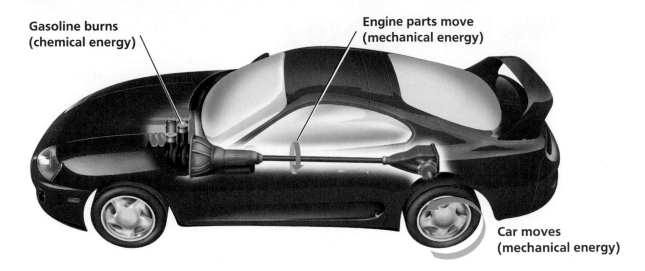

Gasoline burns (chemical energy)

Engine parts move (mechanical energy)

Car moves (mechanical energy)

chemical energy in the battery is changed into electrical energy, heat, and light.

The chemical energy from batteries can change to other forms of energy, too. Think of a toy that runs on batteries. Some batteries cause toy cars to move. When this happens, the chemical energy in the batteries changes to electrical energy, which changes to mechanical energy. Mechanical energy is the energy of motion. The toy car moves across the floor. Other toys can make sounds or "talk" when you put batteries in them.

Imagine how different the world would be if energy couldn't change from one form to another. We all depend on different forms of energy in the world around us every day.

Here is what happens when gasoline burns in a car engine. Mechanical energy is the energy of motion.

3. **Heat makes the wire glow and creates light.**

2. **Electrical energy heats the wire in the lightbulb.**

1. **Chemical energy in the batteries changes to electrical energy.**

An energy change occurs when one form of energy changes into another form. In the flashlight, chemical energy changes to electrical energy and then to heat and light.

How We Use Energy

We use energy in many important ways. We don't use it just to run toys and flashlights. In an electric stove, electrical energy changes to heat. The coils on top of the stove become hot. Heat from the coils is used for cooking food.

Fuel oil is burned in furnaces. Heat is given off when the oil burns. This heat warms air or water that travels through pipes and vents in homes and other buildings, making the buildings warm.

Burning natural gas changes chemical energy to heat. This heat can be used to cook food.

Not all the heat that is given off during energy changes is useful to people. Have you ever felt the top of a computer monitor when the computer was running? It gives off heat. Radios, television, electric alarm clocks, lightbulbs, and even answering machines give off heat energy that is usually not useful.

Back to You

You have seen how substances can react to produce heat. This heat can then change to light or mechanical energy. Many toys and machines work because of these energy changes.

Your body also produces heat. There's an easy way to tell that this is true. Think of the last time you were playing hard with your friends. Maybe you were playing tag or just running around the playground or park. Did you start to get hot and sweat? Sure you did. That's because the movement of your body was changing some of that stored energy into heat. When the air outside is very cold, you can wear a coat, boots, gloves, and a hat to keep your body's heat from escaping to the air around you. When it is hot outside, you can wear light clothing like shorts and T-shirts to help your body release heat.

Your body uses stored energy when you play. The movement of your body changes some of the stored energy into heat.

CHECKPOINT

1. What is stored energy?
2. How does energy change form?
3. How do people use energy?
?. What happens to energy during a chemical reaction?

ACTIVITY

Measuring a Change in Energy

Find Out

Do this activity to learn how stored energy affects the movement of an object.

Process Skills

Observing
Measuring
Communicating

WHAT YOU NEED

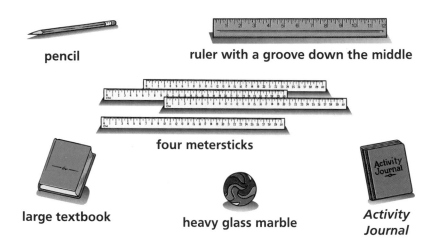

pencil

ruler with a groove down the middle

four metersticks

large textbook

heavy glass marble

Activity Journal

WHAT TO DO

1. Make a channel by laying two of the metersticks side by side on the floor about 2.5 cm apart. Then lay the other two metersticks end to end with the first two. This will make a channel that is 2 m long.

2. Set the ruler flat on the ground with one of its ends against the opening of the channel. Make sure the metersticks are turned so that their lower numbers are near the ruler.

3. Make a ramp by raising the end of the ruler farthest from the metersticks and setting the raised end on the pencil.

4. Place the marble on the raised end of the ruler so that it sits in the groove. Let go of the marble and **observe** its movement.

5. **Measure** how far the marble rolled. **Record** the distance under the heading *Pencil.*

6. Replace the pencil with the textbook. Repeat Step 4.

7. **Measure** the marble's distance. You may have to add the measurement on the second meterstick to the first. **Record** this measurement under the heading *Book.*

8. **Repeat** the activity, again recording all measurements.

CONCLUSIONS

1. Which ramp caused the marble to roll farther, the pencil ramp or the book ramp? Why?

2. At what point in this activity did the marble have the most stored energy?

3. Were your measurements the same the second time you made them?

ASKING NEW QUESTIONS

1. What would happen if you made the ramp even higher by putting two books under the raised end?

2. What sports activities use stored energy to make people and objects move quickly?

SCIENTIFIC METHODS SELF CHECK

✔ Did I **observe** the marble's movement down the ramp?

✔ Did I **measure** the distance that the marble rolled and **record** the measurements?

Review

Reviewing Vocabulary and Concepts

Write the letter of the word that completes each sentence.

1. Matter that is made up of only one kind of atom is an ___.

    a. energy   **b.** evaporation

    c. element   **d.** oxygen

2. Both elements and compounds are ___.

    a. oxygen   **b.** substances

    c. mixtures   **d.** solutions

3. A characteristic of matter is called ___.

    a. a property   **b.** oxygen

    c. energy   **d.** an atom

4. We say that something made of atoms of two or more elements is a ___.

    a. fuel   **b.** compound

    c. heat   **d.** solution

5. When two or more substances combine and each keeps its own properties, we call it a ___.

    a. mixture   **b.** fuel

    c. combination   **d.** solution

Match the definition on the left with the correct term.

6. the change of a liquid into a gas   **a.** chemical change

7. a change that does not result in a new substance   **b.** evaporation

8. a change in which two substances combine and create a new substance   **c.** physical change

9. a substance that we burn to release heat   **d.** battery

10. something that can release energy   **e.** fuel

Understanding What You Learned

1. What do all forms of matter have?

2. List three forms of matter and at least two properties of each form.

3. What do we call water when it is a solid? A gas?

4. What is an example of a substance?

5. What do we call energy that is not being used?

Applying What You Learned

1. What happens to the atoms in a liquid when it evaporates?

2. Explain what happens when an ice cube is changed into water and water is changed into water vapor.

3. What is the difference between a mixture and a solution?

4. Explain how forms of energy change.

 5. What kind of changes can occur when you combine different kinds of matter?

For Your **Portfolio**

Make a poster entitled "Energy Changes Form." From magazines and newspapers, cut out pictures showing people and objects using energy. Under each picture, explain what forms of energy are being used. Draw arrows between different forms of energy to show changes. (Chemical energy→heat)

Light

Light is an important part of our daily lives. In fact, it has been so important to our survival that throughout history many cultures, such as the Egyptians and the Maya, have actually worshiped the sun. Light lets us see the world around us. Light from the sun keeps us warm and helps plants grow to give us the food we need. The sun is the main source of light and other forms of energy here on Earth. In this chapter, we will investigate the many properties of light and the different ways that we use energy from the sun.

The Big IDEA

Light is a form of energy.

CHAPTER SCIENCE INVESTIGATION

Investigate one way that the sun's energy is important to life on Earth by observing the effects of light on plants. Find out how in your *Activity Journal.*

Light Creates Changes

Find Out

- How light makes changes in matter
- How light is converted to heat energy
- How light travels
- How light allows us to see

Vocabulary

solar energy
cornea
pupil
lens
retina

The Big QUESTION

How does light make changes?

The sun is Earth's most important energy source. Our sun is just one of billions of stars in the Milky Way Galaxy, but without its energy we could not survive.

In this lesson, we will explore how light affects plant and animal life, how light travels, and how light and heat are forms of energy that come from the sun. You will also learn how light allows you to see.

How Light Makes Changes

Day after day, the sun produces enough energy to send light out to all the planets in the solar system, including Earth. All life on Earth depends on the energy from the sun. Plants use sunlight to make the food they need to stay alive. People and animals use plants for food and shelter. Plants also produce oxygen with the help of energy from the sun. We all need oxygen from plants in the air we breathe every minute of our lives.

The chemical energy in fireworks is converted to light and heat.

Light and Heat

Another way to explore how light makes changes is to look at the heat that comes from the sun. On a bright sunny day, you can feel the warmth of the sun on your face. Light from the sun warms our planet and

A house with
solar panels

helps cause all weather. The energy we get from the sun is called **solar energy.** The word *solar* refers to the sun. Solar energy warms objects that it strikes because some of the solar energy changes into heat.

Solar energy can also be used to heat homes. To heat a "solar" home, energy from the sun enters glass panels on the roof. Inside the panels are metal pipes with water running through them. The solar energy warms the pipes and the water inside them. This hot water is then pumped through pipes inside the house, warming the entire house. As the water cools down, it is collected in storage tanks. Then, cool water is pumped back up through the pipes to the glass roof panels to be heated up all over again. This is not a form of energy for everyone's house, but a great idea for people who live in sunny climates.

Another way we can use solar energy is to cook meals. The first solar oven can actually be traced back to 1767, when a man from Switzerland named Horace de Saussure (sō′ sur) made a solar oven that cooked fruit. A solar oven is easy to make by covering a surface with aluminum foil, facing it toward the sun, and putting food in a black pot. Because black objects absorb light, the pot takes in heat, and the food inside the pot gets hot.

A Solar Oven

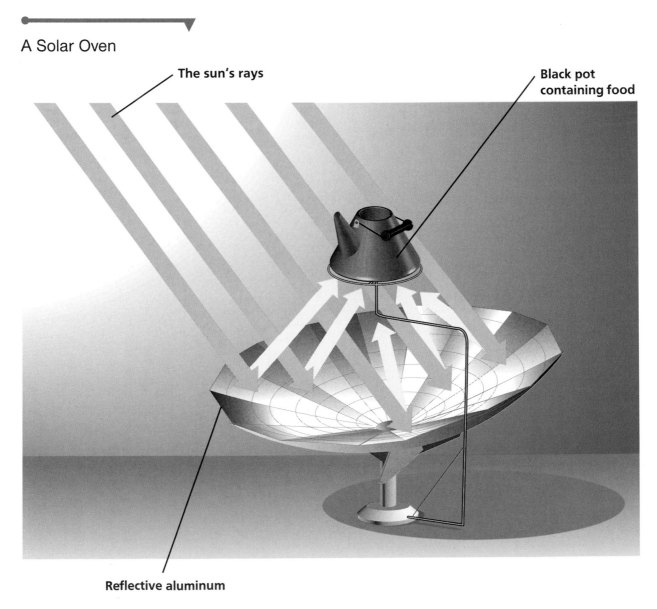

The sun's rays

Black pot containing food

Reflective aluminum

How Light Moves

Have you ever wondered how long sunlight takes to travel from the sun to Earth? The sun is 150 million km away from Earth. Even though the sun is so far away, light travels through space at about 300,000 km per second. This means that light from the sun reaches Earth's surface in about 8 minutes. Nothing else that we know of today travels as fast as light.

How can we find out the direction in which light travels? If you take a flashlight into a dark room and shine it toward a wall, you will see a straight line of light. If you put your hand in front of the light, you stop the light beam because your hand is an opaque object—no light can pass through it. Light travels in a straight line. We also see this when we look at our shadows. If we stand outside on a sunny day facing the sun, a dark shadow appears behind us because the light cannot pass through our bodies. Light can be reflected or blocked by some objects.

Shadows are formed when light is blocked by an opaque object.

The Light You See

We are able to see objects with our eyes because light bounces off of objects. We see objects when light reflected by the object enters our eyes. Without light, we couldn't see anything at all. Light enters our eyes through the **cornea,** a clear "window" at the front of the eye. Then light passes through an opening called the **pupil,** the black spot you see in the middle of your eye when you look in a mirror. In bright light, the pupil becomes smaller, letting less light enter the

We see objects when light reflected by the object enters our eyes.

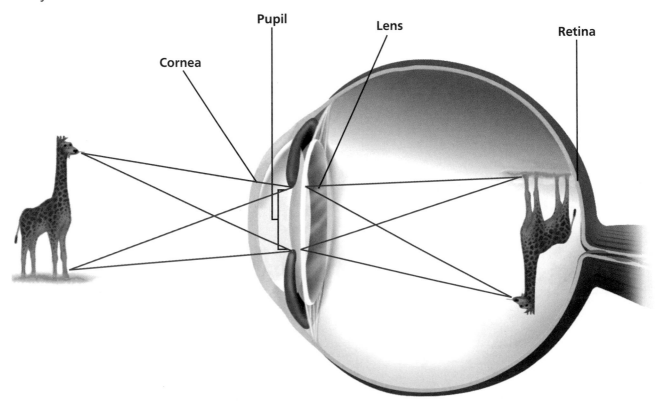

Cornea

Pupil

Lens

Retina

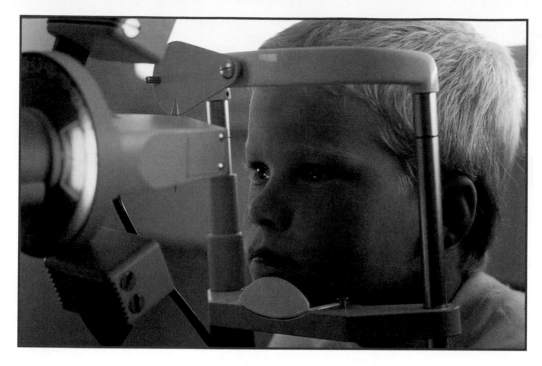

This boy is having his eyes examined.

eye. Behind the pupil is the lens. Both the cornea and lens bend light to form an image of what you are looking at. The **lens** then focuses the image on the **retina** at the back of your eyeball. The image focused on the retina is upside down, but your brain processes the image and turns it right side up as the image you see.

CHECKPOINT

1. How do people and animals use light?
2. How is solar energy useful to people?
3. How do we know that light travels in a straight line?
4. What makes us able to see objects?
 How does light make changes?

ACTIVITY
Making a Window

Find Out

Do this activity to find out which materials allow light to pass through them.

Process Skills

Predicting
Experimenting
Communicating

WHAT YOU NEED

 masking tape

 flashlight

large cardboard box with a 15-cm × 15-cm hole cut in the bottom

18-cm × 18-cm squares of various materials to test: white paper, black paper, wax paper, cloth, aluminum foil, plastic wrap, tissue paper, nylon mesh, and different types of clear plastic

 Activity Journal

WHAT TO DO

1. Set the box on its side on a desk or table so that its bottom is facing away from you. Set the flashlight on another table or desk about one meter behind the bottom of the box. Turn on the flashlight and shine it on the opening in the box. You may want to stack a few books under the flashlight so that it shines directly on the opening in the box.

2. **Observe** each of the test materials. **Predict** which of the materials will allow light to pass through them. **Record** your predictions.

3. One partner should tape a square of paper or other test material over the opening from behind the box. The other partner should stand in front of the box.

 Safety! *Do not look directly at the flashlight beam.*

4. The partner behind the box should hold his or her hand up to the opening in the box, between the opening and the flashlight. The other partner should **observe** and **record** what can be seen through the box.

5. Repeat this step with several of the test materials, one partner removing each square of material from the box and then taping a new one over the opening.

6. Trade places with your partner and repeat the activity, using the rest of the test materials.

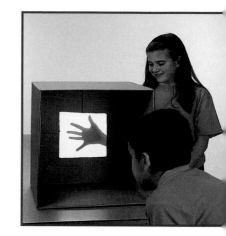

CONCLUSIONS

1. Which of the materials blocked the light?
2. Were your predictions correct?

ASKING NEW QUESTIONS

1. How do objects that block light and objects that do not block light help people?

SCIENTIFIC METHODS SELF CHECK

✔ Did I **predict** which materials would allow light to pass through them?

✔ Did I **experiment** by testing different materials?

✔ Did I **observe** what happened with each material?

✔ Did I **record** my predictions and observations?

Properties of Light

Find Out

- How light travels
- How shadows are made
- How light is reflected

Vocabulary

transparent
translucent
opaque
mirror
scatter

The Big QUESTION

How does light behave?

So far, we have seen that light travels in a straight line. Now it is time to explore what happens when light strikes different surfaces. Surfaces can make light change direction.

Light Movement

When light hits different types of material, different things can happen. If we shine a light behind a glass of water, the light will shine through because both the glass and the water are transparent. **Transparent** objects allow light to pass through them clearly.

Light reflecting off
buildings

Light reflecting off
ripples on a lake

Transparent

Translucent

If we added a little milk to that glass of water and tried to shine a light through, what do you think would happen? Since the milk is not transparent, it will mix with the clear water to make a liquid that allows only some of the light to shine through. Matter that lets only some light pass through it is **translucent.** The translucent milk-water mixture will scatter or separate the light rays, and only some of the light will travel through the water, milk, and glass. Finally, if we poured this liquid into a blue plastic mug we would see that the light would not be able to pass through the object. The plastic mug is **opaque.**

Opaque

12:00 noon

5:00 PM

Shadows

Have you ever played shadow tag or made shadows on the wall? Do you remember how you made the shadows? A shadow is created when an opaque object blocks a light source. Shadows appear to change shapes and sizes. We can see shapes and shadows change with the movement of the sun throughout the day. If the sun is overhead, many shadows will look short. In the morning as the sun rises or in the evening as the sun starts to set and gets lower in the sky, these shadows look longer.

Reflections and Mirrors

Light rays bounce off smooth surfaces in only one direction.

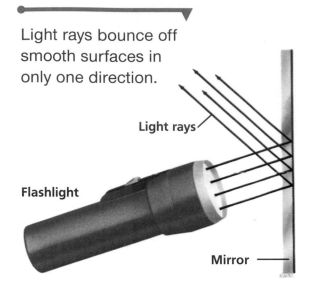

Light rays

Flashlight

Mirror

When we look in the **mirror,** we are seeing our image. Since a flat mirror has a shiny, flat surface, it reflects a clear image. Light rays bounce off the mirror in a regular way, allowing us to see the mirror's image clearly. For example, light bounces off a flat mirror like a tennis ball hitting the floor. What do you think might happen if light rays hit a rough surface like a sidewalk or a carpet? Since the surface of the object is not smooth, the light rays scatter and create no image. To **scatter** is to separate and move in many different directions. This is because the light rays bounce off in many

Light rays bounce off rough surfaces in many directions, causing the light rays to scatter.

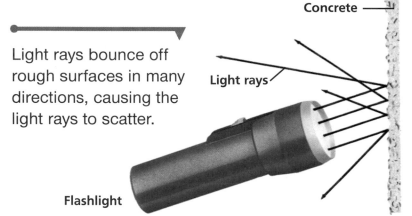

Concrete

Light rays

Flashlight

You can see your reflection in a mirror because light rays bounce off the smooth surface in only one direction.

Although this woman can see a wider area in the image, things in the image seem smaller.

different directions. So, you may be able to see your reflection in a store window or a shiny floor, but not on a brick wall or on the sand at the beach!

Mirrors are not always flat. Sometimes they are curved. Have you ever looked at the mirror on the passenger side of a car? If you have, you probably saw a mirror that curved outward. This type of mirror reflects light from a wider area, allowing drivers to see more of the traffic behind them.

CHECKPOINT

1. How does light travel?
2. When can you see a shadow?
3. Describe what happens when light hits a flat, shiny surface. What happens when light hits a rough surface?

 How does light behave?

ACTIVITY

Bouncing Light

Find Out

Do this activity to find out how light reflects.

Process Skills

Observing
Experimenting
Communicating

WHAT YOU NEED

textbook

flashlight

three mirrors

Activity Journal

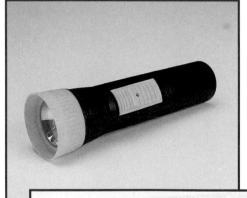

WHAT TO DO

1. Place a textbook so it stands with the spine side up. Lean a mirror against the textbook.

2. **Write** your name or a word on a piece of paper and hold it up to the mirror. **Observe** how your name appears in the mirror.

3. **Look** at your face in the mirror and touch the left side of your face. Which side of your face did your reflection touch?

4. Place the textbook in the middle of your desk and point a flashlight directly at the front cover.

5. Without moving the book, **experiment** by using all three mirrors to try to get the flashlight beam to show up on the other side of the book. Try doing this many different ways.

6. **Draw** a diagram of how you placed the mirrors to make this happen.

CONCLUSIONS

1. What did the images look like in the mirror?

2. In what ways did you try to make the beam of light hit the other side of the textbook?

3. What was happening to make the light bounce?

ASKING NEW QUESTIONS

1. What do you think would happen if you wrote a word backward and then held the paper up to a mirror? **Hint:** Print the letters backward, too.

2. You look in a mirror in the morning to brush your hair. Where else do people use mirrors?

SCIENTIFIC METHODS SELF CHECK

✔ Did I **observe** the images in the mirror?

✔ Did I **experiment** with the bouncing light on the book?

✔ Did I communicate by **drawing** a diagram of the experiment?

The Spectrum of Light

Find Out

- How a prism separates light
- What a spectrum is
- How we see colors

Vocabulary

prism
refract
spectrum

The Big QUESTION

What are the named colors in the spectrum of light?

*I*n spring, we notice the deep green of the grass. In summer, we are aware of the bright blue of the sky. In the fall, we look at the bright reds and oranges of the turning leaves. Colors are all around us. Have you ever thought about how we see colors?

In this lesson, we will look at how light interacts with the objects around us to better understand the relationship between light and color.

Rainbows show all of the colors that make up visible light.

Prisms

Although sunlight might seem to be white, it is really a mix of many colors. To see these colors we can look at light that has passed through a prism. A **prism** (priz′ əm) is a piece of glass or plastic that can **refract,** or bend, light. When white light goes through a prism, the light separates into a rainbow of colors. For example, after a rainy day we are able to see a rainbow when the sun comes out. When this happens, small drops of water in the air act like prisms to refract the sunlight. You may have seen how a prism can separate colors by looking at crystals that people sometimes hang in their windows to catch the sunlight.

A prism

A miniature rainbow is formed by a prism. A prism is a piece of glass that's cut and polished, like crystals you might see people hang in their windows to catch the sunlight.

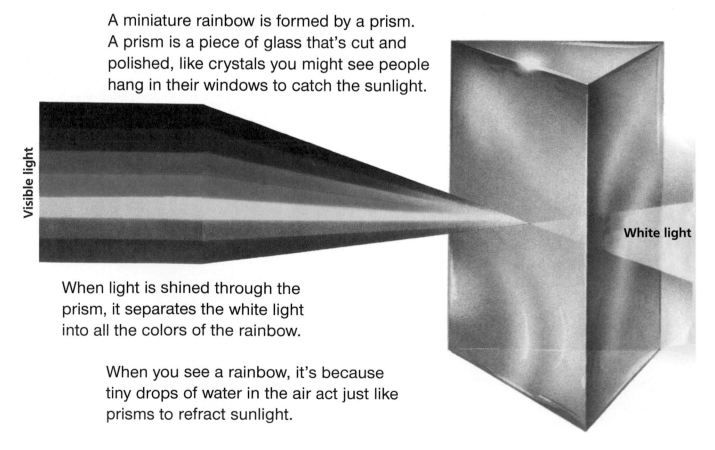

Visible light

White light

When light is shined through the prism, it separates the white light into all the colors of the rainbow.

When you see a rainbow, it's because tiny drops of water in the air act just like prisms to refract sunlight.

The Light Spectrum

The Spectrum

This rainbow of color is called a light spectrum. The **spectrum** shows all of the colors that make up visible light. Although a spectrum has a broad range of colors, people usually name seven of the colors in order—red, orange, yellow, green, blue, indigo (dark blue), and violet. To make it easier to remember, think of the colors by their initials, ROY G BIV.

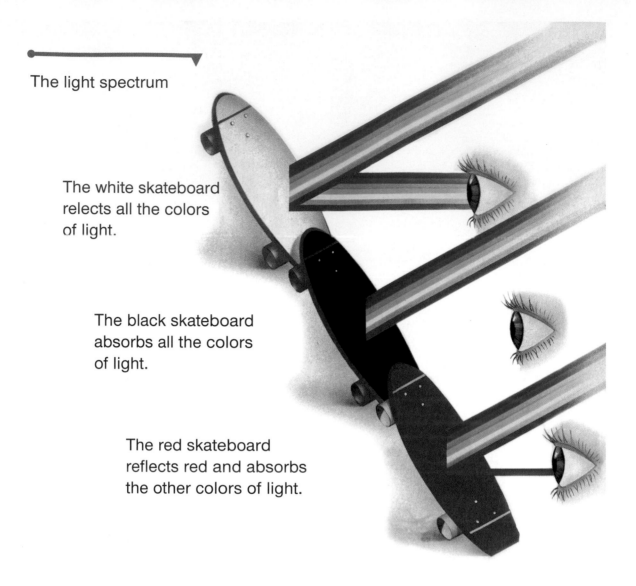

The light spectrum

The white skateboard relects all the colors of light.

The black skateboard absorbs all the colors of light.

The red skateboard reflects red and absorbs the other colors of light.

How We See Color

You see colors because of the way objects either reflect or absorb colors of the light spectrum. A green object looks green because it absorbs all of the colors of the spectrum except green. It reflects green light, so the color of light that reaches our eyes is green. So, when we look at the green grass, only the green light is reflected to our eyes. The grass has absorbed all of the other colors in the spectrum.

The sky looks blue on sunny days.

Painting with different colors

Do you know why the sky looks blue on a sunny day? As sunlight shines through Earth's atmosphere, the blue light from the spectrum scatters in all directions. Other colors do not scatter as much. When you look up at the sky on a sunny day, this blue light shines into your eyes. So, even though the light that comes to Earth from the sun includes all colors, blue is the color that our eyes often see in the sky.

CHECKPOINT

1. What is a prism and how does it work?
2. What is the light spectrum?
3. How does light affect how we see colors?
 What are the named colors in the spectrum of light?

ACTIVITY
Seeing White Light

Find Out

Do this activity to learn how white light and other colors are created.

Process Skills

Observing
Experimenting
Communicating

WHAT YOU NEED

transparent tape

three flashlights

four pieces of construction paper cut into 30-cm squares (white, red, blue, green)

three pieces of colored cellophane to cover the flashlights (red, blue, green)

Activity Journal

WHAT TO DO

1. Tape a piece of cellophane over each flashlight head.

2. Turn out the lights. Stand close together with two partners and focus the three flashlights on the white sheet of paper at the same place. **Observe** the light.

3. **Record** what you see.

4. Next, shine your red flashlight on the blue paper. Shine the green flashlight on the red paper. **Observe** what happens.

5. **Record** what you see.

6. **Experiment** by shining two of the flashlights on one of the colored pieces of paper until the two lights overlap or meet at the same place.

7. **Record** your observations.

Conclusions

1. What happens when you shine all three flashlights in the same place?

2. What happens when you shine two different colors together on the white paper?

3. What happens when you shine different colored lights on the different colored papers?

Asking New Questions

1. What color would a red rose appear to be if it were lit by a blue light?

2. Do you think it makes a difference what kind of surface you shine your lights on?

SCIENTIFIC METHODS SELF CHECK

✔ Did I **observe** what happened when I **experimented** by using different colors of light on different colors of paper?

✔ Did I **record** my observations?

Review

Reviewing Vocabulary and Concepts

Write the letter of the word or phrase that completes each sentence.

1. The part of the eye through which light passes is ___.
   a. the pupil   b. the retina
   c. a solar panel   d. an image

2. The part of the eye that bends light and focuses what you see is the ___.
   a. cornea   b. eyelid
   c. prism   d. lens

3. Objects that let light pass through clearly are ___.
   a. opaque   b. transparent
   c. plastic   d. visible

4. Material that lets only some light pass through it is ___.
   a. clear   b. solar
   c. curved   d. translucent

5. You cannot see your reflection in a rough surface because the light rays are ___.
   a. reflected   b. bent
   c. scattered   d. opaque

Match each definition on the left with the correct term.

6. energy we get from the sun   a. prism

7. part of your eye at the back of the eyeball   b. solar energy

8. an object that separates light   c. refract

9. to bend light   d. spectrum

10. the colors of visible light   e. retina

Understanding What You Learned

1. What is the source of light on Earth?

2. How does light help us live on Earth?

3. What about light allows us to make shadows?

4. How can a mirror show an image?

5. Why do things in the world have different colors?

Applying What You Learned

1. Name some things in your classroom or home that are transparent.

2. Why is the world more colorful on a sunny day than on a cloudy day or in the evening?

3. List some other sources of light people use.

4. Which colors of the spectrum are reflected by a blank sheet of white paper?

 5. How do living things use light as a form of energy?

For Your **Portfolio**

Draw pictures of animals, plants, or other objects that are the colors of the spectrum. Categorize the things you draw according to color. Then make a "rainbow" display of all the things you drew.

Simple Machines and

Every day we use machines. Think about something as common as the food we eat. We use machines on farms to plant and harvest crops. We use machines to transport food to markets. You fill your cart with things you want to buy. At the checkout counter, the prices of your groceries are totaled. There may be a moving belt that carries your food to the bagger. Then, you have to get it home and put it away. How many machines did it take just for you to eat breakfast this morning? We are going to learn about how some simple machines help us do work.

The Big IDEA

Simple machines help us to do work.

How They Work

CHAPTER SCIENCE INVESTIGATION

Build a model of a simple machine to observe how distance, force, and work are related. Find out how in your *Activity Journal.*

GAS RANGE

Making your world a little easier

2059

Force and Work

Find Out

- What force is
- How work is done
- How force can be measured

Vocabulary

force
gravity
speed
work
distance

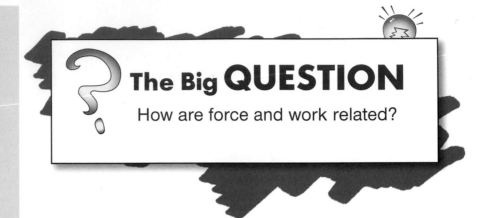

The Big QUESTION

How are force and work related?

Every day you experience several different forces, but you may not call them forces. You may say, "This book was pushed across the desk," or "I pulled my brother to school in a wagon."

Force

Force is the effort made when matter is pushed or pulled. Forces can cause objects to change their motions. All pushes and pulls are forces. You get up in the morning by pulling or pushing yourself out of bed. You use force to open or close a door, lift a book, and throw a ball.

To move, objects must have a force exerted on them. To exert a force is to use effort to make that force happen. For example, when you ride a bicycle, you exert force on the pedals with your feet and legs. This force makes the pedals move, which causes the gears and wheels of the bicycle to move. Your legs and feet, by pushing on the pedals, exert the force that moves the bicycle.

Energy is used to exert force. Energy is the ability to make things move or change. Your muscles use the food you eat to get energy to move objects. Electric motors get energy from electricity to cause movement. The motor in a car gets energy from burning gasoline to cause the car to move. The energy of moving air makes kites and pinwheels move. Even though the air is invisible, it has a lot of force.

Magnets can exert force on iron objects, even when they are not touching. You can observe this if you put a paper clip a centimeter away from a magnet. The clip will be attracted to the magnet when you let it go. Magnets can also push or pull each other. If you turn two magnets toward one another and gently push them together, you will feel the forces exerted by the magnets that keep them apart.

Everyday activities involve pushes and pulls.

How many pushes and pulls do you see in this picture? Who is pushing? Who is pulling? Who is doing both?

When you lean (push) against something, the thing you are leaning against is pushing back with an equal amount of force.

Another force is pulling on you right now. Can you guess what it might be? The force of **gravity** pulls objects toward Earth. That is the reason things fall down when they are not supported. It also keeps the objects on Earth from drifting off into space.

Motion

Motion is what happens when something changes position. All motion takes time to happen. **Speed** is the measure of how fast an object moves over a distance. Some things change position in very little time. When you throw a ball to a classmate, it gets there in just a few seconds. Some things move so fast you don't even see them move, like the wings of a hummingbird in flight.

Some things move so slowly that they seem to be standing still. Look at the hands of a clock. Can you see the hour hand move? How can you

Hummingbirds' wings move so fast in flight that they appear as a blur.

When you ride a bicycle, you exert force on the pedals with your feet and legs.

tell that it moves? Where is the hour hand when you start school? Where is it at lunch time? By recording the change in position, you can see that the hand does move.

No matter how quickly or slowly an object moves, all moving objects are changing position. Forces cause objects to change position. When a leaf falls from a tree, the force of gravity causes the leaf to move to the ground. When you kick a soccer ball, you exert the force that makes the ball roll along the ground or soar through the air.

To make a ball move, you use your body's energy to exert a force. That force allows you to move the ball over a short distance or a long distance, depending on how much force you exert.

Work

Work is done when a force makes an object move over a distance. **Distance** is how far something moves. When you push a cart at the grocery store and it moves, you do work on the cart. When a can of peas falls off the shelf, gravity does work on the can, making it crash to the floor. However, if you push against a large rock and it does not move, you have done no work on the rock. No matter how hard you push or pull or how tired you get, until the rock moves, there is no work done on it.

Measuring Force

Scientists find it useful to measure the forces around us. You probably already know some ways of measuring force. Stepping on a scale tells you how much you weigh. Weight is the way we measure the pull of gravity. A spring scale can be used to measure the amount of force of a sideways pull. Think about all the different kinds of scales there are.

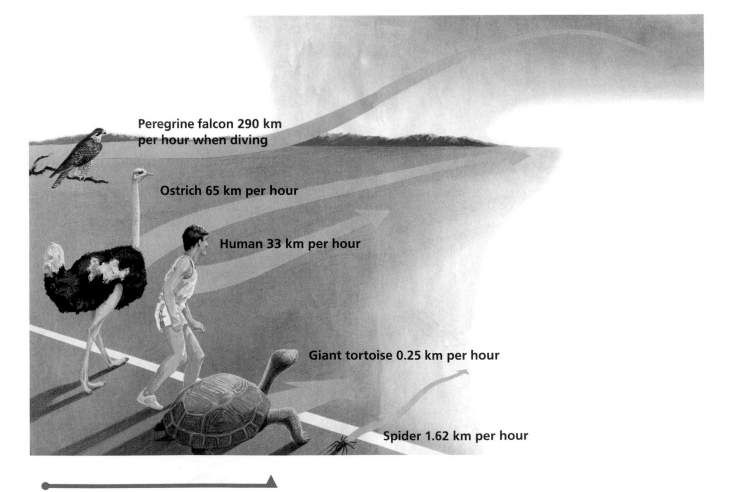

Peregrine falcon 290 km per hour when diving

Ostrich 65 km per hour

Human 33 km per hour

Giant tortoise 0.25 km per hour

Spider 1.62 km per hour

Look at the top speed for each animal above. Which one would win a race?

To measure how fast an object is moving, you measure the object's speed. If something is moving, its position is changing. You can tell that it is moving by comparing its position to something that is not moving. You can compare how fast things move by finding their speeds. Speed is easily measured by recording both the distance that an object has moved and the amount of time that it took to move the distance, and then dividing the distance by the time. For example, if you walked 5 km in one hour, your speed was 5 km per hour.

There are many ways to measure speed. To know the speed of an object, you need to know how far it has moved and how long it took to get there. In the United States, we usually measure speed in miles per hour. Other countries use kilometers per hour. If you wanted to measure smaller distances you could use different units, such as feet per minute.

CHECKPOINT
1. What is force?
2. What does work do to objects?
3. What two things must you know to measure speed?
 How are force and work related?

ACTIVITY
Measuring Force

Find Out

Do this activity to find out how to measure the force it takes to move common objects.

Process Skills

Observing
Measuring
Communicating

WHAT YOU NEED

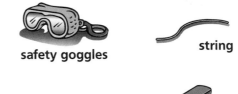

safety goggles   string

spring scale

objects to weigh: box of crayons, tape dispenser, stapler, lunch box, and so on

Activity Journal

WHAT TO DO

1. Hook your finger around the bottom of the spring scale and *gently* pull while your partner holds the top of the scale. Make the scale read 10 N. This reading on the scale means that your pull has a force of 10 newtons. A newton is a unit used to measure pushes and pulls. Does it matter how either of you pulls? **Observe** what happens. **Record** your observations.

2. Now hold the bottom of the spring scale while your partner *gently* pushes away from the top of the scale. Make the scale read 10 N again. **Record** your observations.

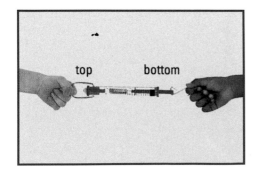

top   bottom

3. Now, **measure** the amount of force needed to move each of the objects you have collected for the activity. Hook the scale on an object. Place the scale and the object on a smooth, flat surface and pull at a steady speed. **Record** what the spring scale reads. Then have your partner hold the scale while you *gently* push the object away from the scale. **Record** the reading. Repeat this step with each of the other objects. **Record** each of the readings for both pushes and pulls.

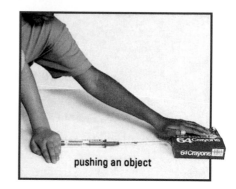

pushing an object

CONCLUSIONS

1. What did you feel when your partner pushed or pulled on the spring scale?

2. Which objects made the scale read the highest when you pulled and pushed them?

3. Why did it take a bigger pull to move some objects?

ASKING NEW QUESTIONS

1. Both pushing and pulling made the spring scale read 10 N. How are pushing and pulling the same?

2. Why is it useful to be able to measure force?

SCIENTIFIC METHODS SELF CHECK

✔ Did I **observe** the effects of pushes and pulls on the reading of the spring scale?

✔ Did I **measure** the forces needed to push and pull the different objects?

✔ Did I **record** the measurements and my observations?

Simple Machines

Find Out

- How an inclined plane works
- How levers work
- How other simple machines work

Vocabulary

inclined plane
lever
fulcrum
load
effort

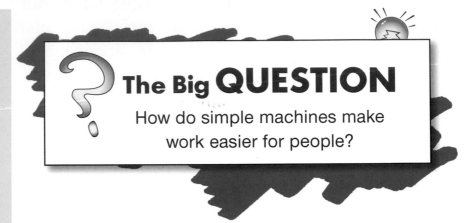

The Big QUESTION

How do simple machines make work easier for people?

Every day, heavy objects are moved from one place to another. Trucks carry bricks and lumber. Cranes move material around a work site. Grocery stores fill their shelves with food. Packages are delivered. Imagine how tired we would get if all these things had to be carried by hand from place to place! Instead, we use machines to make it easier to move things. Trucks and cranes are complicated machines with many moving parts. Engines burn fuel to give them energy to move their loads. Many simple machines are also used every day to make it easier to move objects.

Inclined Planes

One simple machine that you have used many times is an inclined plane. An **inclined plane** is a flat surface that is higher at one end than at the other. A slide is an inclined plane. So are the wheelchair ramps that you see outside many buildings. Moving up an inclined plane uses less force than moving up stairs or up a ladder. When walking up an inclined plane, you can push or pull a heavier object than you would be able to carry up stairs. Can you think of times when you have used an inclined plane? When using an inclined plane, you move a longer distance but you use less force. Changing the slant of the inclined plane changes the amount of force needed to reach the top.

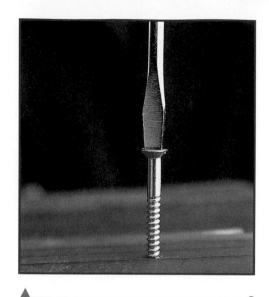

Simple machines help us do many different kinds of work. A screwdriver is a simple machine and a screw is an inclined plane.

Using an inclined plane in ancient Egypt

A slide is an inclined plane.

Three Kinds of Levers

Another simple machine that you have probably used is a lever. A **lever** is a flat surface or other piece of strong material that rests or tilts on a **fulcrum,** or pivot point. Levers work when a lifting force is exerted at some point along their length. Levers allow people to move objects that are too heavy to lift easily.

One kind of lever has its fulcrum in the middle. At one end is a **load,** or item to be moved. At the other end, you apply an effort or force. The **effort** is the push or pull that gives the energy to move the load. A crowbar or other kind of pry bar is an example of a lever. A crowbar is used for prying apart or lifting materials. One end of the bar is placed under the load, the center of the bar is the fulcrum, and the effort is applied at the other end. Scissors and balance scales are examples of levers that have the fulcrum in the middle.

A crowbar is an example of a lever that has the fulcrum in the middle.

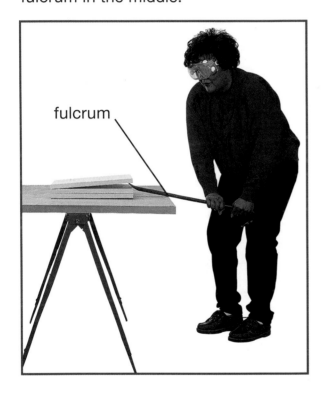

fulcrum

Scissors are also levers. The handles are where the effort is applied, the center (pivot point) is the fulcrum, and the blades are the load.

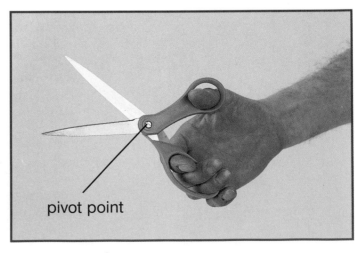

pivot point

A wheelbarrow is an example of a second kind of lever. The wheel is the fulcrum and the effort is the force you exert when you lift the handles. The load is in the middle. With these levers, the effort travels a longer distance than the load. This makes it easier to lift a heavy object. A nutcracker is also a lever. The hinge is the fulcrum. The load is the force it takes to crack the nut, which goes in the middle. The force is applied at the end of the handles. We often think of levers as lifting an object, but this is one example where the levers squeeze an object.

The fulcrum of a third kind of lever is at one end of the lever, and the load is at the other end. The effort is applied in the middle. These levers use a larger effort to move the load a greater distance. Think of a broom. You hold it with one hand on top and the other hand in the middle to apply the effort. The top of the broom is the fulcrum. It stays in one place, the middle moves a little, but the bottom moves a lot as it sweeps across the floor, cleaning up a mess. A canoe paddle is also a lever. It also uses a greater effort to make its load move more quickly.

A canoe paddle is an example of a lever.

Other Simple Machines

There are other simple machines that make work easier. A wedge will split objects apart.

Another kind of lever is a wheel and axle (ak′səl). A wheel and axle is a simple machine with a wheel that turns a post. The post is called an axle. Many common objects use a wheel and axle. A doorknob pulls the latch with a wheel and axle. How many wheels do you see in this picture? Which ones are part of a wheel and axle?

A third kind of lever is a pulley (pül′ ē). A pulley is a simple machine in which a rope fits around a wheel. A pulley is often used to lift things. It lets you stand on the ground and pull a rope down to lift something up. Where is the pulley in this picture?

It's easier to lift a very heavy object or pry something up with a lever than without one. Tools that can pry or pull things up are levers. Where is the lever that is prying something up in this picture?

A compound machine is made up of two or more simple machines that are connected. Two of the simple machines in this picture are working together as a compound machine. Can you find it?

A screw can pull objects together. Wheels and pulleys transfer force to move objects sideways, up, or down.

Machines can't do work by themselves. Machines do work by transferring energy. Energy is the ability to do work or make something move by exerting a force. A machine cannot transfer any more energy than you put into it. Many machines depend on your muscle power to make them move. Others get power from gravity, electricity, and even magnetism.

A knife is a simple machine.

CHECKPOINT

1. How does an inclined plane make it easier to lift an object?

2. Give an example of each of the three kinds of levers.

3. What are some examples of simple machines?

 How do simple machines make work easier for people?

ACTIVITY
Making a Lever

Find Out

Do this activity to find out how changing the distance a lever moves will change the force needed to do work.

Process Skills

Constructing Models
Observing
Communicating

WHAT YOU NEED

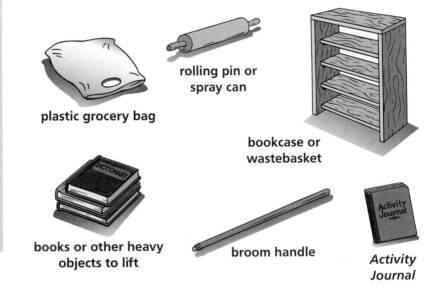

plastic grocery bag

rolling pin or spray can

bookcase or wastebasket

books or other heavy objects to lift

broom handle

Activity Journal

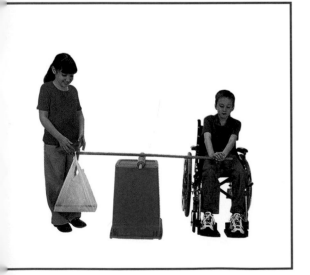

WHAT TO DO

1. Place the rolling pin (the fulcrum) crosswise under the center of the broom handle (the lever). Rest the fulcrum on top of the bookcase or an upside-down wastebasket. Put the weight in the plastic bag and put it on one end of the pole. Push on the other end to lift the load. This is your model lever.

2. Move closer to the fulcrum and push down. **Observe** what happens. What changed? Move the weight closer to the fulcrum. Push on the end of the lever. **Record** your observations.

3. Make a second kind of lever by placing the fulcrum under one end of the lever. Rest the fulcrum on the bookcase. Put the load halfway along the lever. Have a partner hold the fulcrum down while you lift the other end of the lever. **Observe** what happens. How far did your hands move? How far did the load move? Trade places and do it again. **Record** your observations.

4. Move the load to the end of the lever. Hold down the fulcrum while your partner lifts the lever in the center. This is a third kind of lever. Which moves farther—your hands or the load? **Record** your observations.

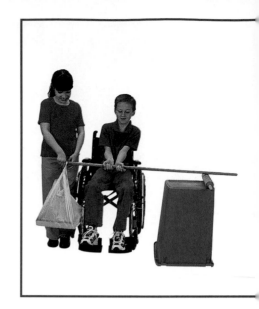

CONCLUSIONS

1. Which kind of lever would you use to move a very heavy object? Why?

2. Which kind of lever would you use to lift an object high in the air? Why?

ASKING NEW QUESTIONS

1. Look around at school and at home to see levers in use. Can you find at least five?

2. What are some everyday uses for the kinds of levers you made in this activity?

SCIENTIFIC METHODS SELF CHECK

✔ Did I **make a model** of each kind of lever?

✔ Did I **observe** and **record** what happened as I changed the position of the load, the fulcrum, and the force?

Review

Reviewing Vocabulary and Concepts

Write the letter of the word or phrase that completes each sentence.

1. The force of ___ pulls objects toward Earth.
   a. energy   **b.** light
   c. gravity   **d.** work

2. ___ is how far something moves.
   a. A meter   **b.** Distance
   c. A kilometer   **d.** A centimeter

3. ___ is a flat surface that is higher at one end than at the other.
   a. A table   **b.** A floor
   c. A road   **d.** An inclined plane

4. ___ are simple machines that allow people to move objects that are too heavy to lift easily.
   a. Trucks   **b.** Levers
   c. Loads   **d.** Rocks

5. The push or pull that gives the energy to move the load is ___.
   a. effort   **b.** strain   **c.** energy   **d.** work

Match each definition with the correct term.

6. the effort made when an object is pushed or pulled   **a.** fulcrum

7. the measure of how fast an object moves over a distance   **b.** work

8. a force making an object move over a distance   **c.** force

9. the pivot point for a lever   **d.** load

10. an item that is to be moved   **e.** speed

Understanding What You Learned

1. How is force measured?

2. How is work done? Give an example.

3. Why is it easier to carry a heavy object up an inclined plane than up stairs?

4. Name three kinds of levers.

5. Explain the parts of a lever.

Applying What You Learned

1. What makes the pedals move when you ride a bicycle? What happens if you stand up and pedal?

2. How are force and motion related?

3. If you had to move a very heavy object up to a higher place, what simple machine would you use? Why?

4. Where is the fulcrum on a pair of scissors?

5. How do simple machines help us do work?

For Your Portfolio

Write a short story that features a simple machine. Choose a simple machine that can help your character(s) with a problem. For example, a character could use a lever to lift a heavy box off a floor, a simple pulley to move a bucket to a tree house, or a ramp to help a friend who uses a wheelchair into a home.

ELECTRICITY

Energy makes things happen. It makes it possible for everything—including you—to move around and to do work. Living things get their energy from food. Your body stores the energy you get from food. Then, your body releases that energy in actions like getting dressed, running, riding your bike, or thinking.

In a lightbulb or toaster, electrical energy is changed into light and heat. You have learned about energy being changed to heat and light. Another form of energy is electrical. Long ago, people used only their own energy to work. Today, we use electrical energy to do a lot of work. It has made our lives much easier.

The Big IDEA

Electricity is a form of energy.

CHAPTER SCIENCE INVESTIGATION

You can make a game using electrical circuits. Find out how in your *Activity Journal.*

Electrical Energy

Find Out

- What electrical energy is
- How electrical energy is made and moved
- How electrical energy does useful work

Vocabulary

electrical energy
circuit
electric current
conductors
nonconductors

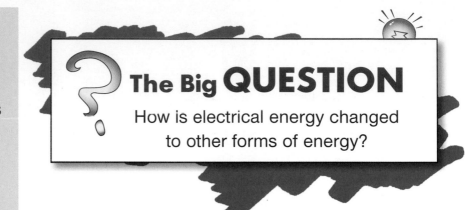

The Big QUESTION

How is electrical energy changed to other forms of energy?

Electrical energy! What came to mind when you read those words? You might have thought of the lights in your home. Maybe you thought about your television or video games. Or maybe you pictured a power plant or huge power line towers near your home. We use electrical energy to do many things. This lesson explains what electrical energy is and how it is made. It also tells about how electrical energy is changed to other forms of energy that are useful to people.

What Electrical Energy Is

The energy of electric current flowing through wires can make whatever is connected to it work. This energy is called **electrical energy.** Because it is energy, it can be changed to other kinds of energy. The electrical energy that flows through wires into your home is changed into heat, light, or mechanical energy. It may be changed into heat energy in an oven, light in lightbulbs, or mechanical energy to run the motor of a dishwasher.

Electrical energy is part of our daily activities.

Making and Moving Electrical Energy

A Power Source

Other kinds of energy are needed to make electrical energy. Power plants change other kinds of energy into the electrical energy that flows through wires to your home. Power plants have huge machines called *generators* that use other forms of energy to make electrical energy. Some power plants use the energy of running water to run their generators. The water is held behind a dam until it is needed to generate electrical energy. Other plants use energy from burning coal, oil, or gas to run their generators. Others use nuclear energy.

Batteries also produce electrical energy. Batteries have chemicals in them that react to release electrical energy. You may have used batteries to power a flashlight, a radio, or a calculator.

A dam and power plant use the energy of moving water to generate electricity.

A Pathway to Follow

Electrical energy needs a pathway to follow to get from one place to another. The pathway that electric current moves through is called a **circuit** (sûr′ kət). The electrical energy that flows through a circuit is called **electric current.** It is called a current because electrical energy flows through wire circuits just as currents of water flow through a river channel. The big wires you see strung on top of tall poles are part of a circuit. They carry electric current from the power plant to your home.

A circuit must be continuous like an unbroken circle. If a circuit wire is broken or cut, electric current can't get through. Sometimes wires from a power plant get damaged by storms or cut by accident. When that happens, no electrical energy gets to your home. The circuit has been broken and the lights go out.

The wires in your home are covered with materials like plastic that don't conduct electricity. These materials help keep you safe when you use electrical appliances.

Electric current flows through some materials better than others. Materials that electric current flows through easily are called **conductors** (kən duk′ tərs). Metals like copper and aluminum are good conductors. Wires make good conductors because electric current flows through the metal in the wires very easily.

Your body conducts electric current, too. That is why it is dangerous to stick your fingers into electrical outlets. Electric current can flow through your body and cause serious burns or death.

These materials are good conductors of electric current.

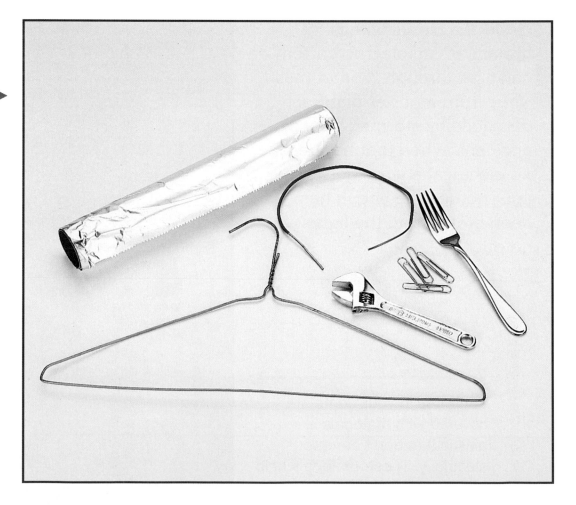

These materials do not conduct electric current very well.

Some materials don't conduct electric current very well. These materials are called **nonconductors** (non′ kən duk′ tərz). Most plastic and rubber materials are nonconductors. All the wires around your home are covered with materials like plastic or rubber that don't conduct electricity. They keep wires from touching each other or other things. Covering the wires also keeps people from touching wires by mistake and getting electrocuted.

Putting Electrical Energy to Work

To be useful, electrical energy needs to be changed to other kinds of energy. In your home, electrical energy is often changed to light, heat, or mechanical energy.

Imagine that you want to make the lightbulb in your bedroom light up. First, the power plant makes electrical energy from another form of energy. The electric current flows through wires into your home. Then, a smaller circuit branches off the main circuit in your home and goes to switches in your bedroom.

When you turn the light switch on, you start the electric current flowing. The current heats up a very thin wire in the lightbulb. The wire gets so hot that it glows with white light. Electrical energy has been changed to light in your lightbulb.

People use electrical energy for many other things in their homes. It is changed to heat by electric heaters, electric ranges, or irons. It is changed to mechanical energy to run motors in washing machines, clothes dryers, fans, or computers. It is changed to sound in telephones and radios.

Batteries also generate electrical energy. Batteries power portable radios, flashlights, calculators, watches, and games. Batteries are useful because they let you use something without being connected to the electric current in your home. Batteries do

In a dishwasher, electrical energy is changed to mechanical energy.

Electric cars run on electrical energy from rechargeable batteries instead of on gasoline.

stop making electrical energy after they have been used for a while. This happens when all of the chemicals inside the battery have reacted with each other. Some batteries are rechargeable. The battery in a car is rechargeable. To recharge a battery, electrical energy is used to change the battery's chemicals back to the original substances that the battery had when it was new.

CHECKPOINT

1. What is electrical energy?
2. How is electrical energy made and moved?
3. How is electrical energy used in our homes?
4. How is electrical energy changed to other forms of energy?

ACTIVITY

Making a Squeeze Circuit

Find Out

Do this activity to find out how to make a model of the path of electric current traveling through a circuit.

Process Skills

Constructing Models
Communicating

WHAT YOU NEED

Bulb
one label
saying "Bulb"

tape

D-cell
one label
saying "D-cell"

Activity Journal

WHAT TO DO

1. Stand in a circle with five partners, facing one another. Label one student "D-cell" and a second student "Bulb." The four other students will act as "wires."

2. Figure out how the six of you must be arranged, holding hands, to act out the path of electric current flowing through wires. **Make a model** of the path by arranging your group in the order that the current must flow. The current must end up lighting the "bulb."

3. Because the current must begin at the "D-cell," the student with the "D-cell" label should act out the beginning of the flow of electric current by squeezing the hand of one of the "wires."

4. The "wire" should then squeeze his or her other hand to "pass the current" on to the next "wire." Repeat this process with each of the "wires" until the "bulb" has been lighted. Then, continue until the "current" (squeeze) is returned to the "D-cell."

5. **Record** what you and your classmates did to model the path of electric current by drawing a picture of it. Label each part of your drawing to show the D-cell, the bulb, and each of the wires.

CONCLUSIONS

1. How did the students in your group have to arrange themselves to act out the flow of electric current from a battery to a lightbulb?

2. Why was this arrangement so important to show the flow of electric current?

ASKING NEW QUESTIONS

1. What would have happened to the flow of electric current in your model if one of the "wires" had let go of the hand of another "wire"?

SCIENTIFIC METHODS SELF CHECK

✔ Did I **make a model** of an electric circuit with my classmates?

✔ Did I **record** the arrangement of the students in the model by drawing a picture of the model?

Electromagnets

Find Out

- What magnetism is
- How electric current can make a magnet
- How we use electromagnets

Vocabulary

magnetism
electromagnet

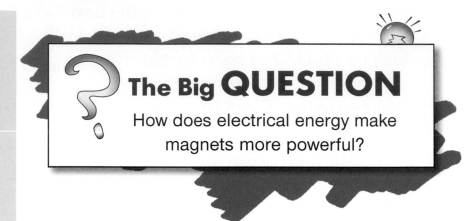

The Big QUESTION

How does electrical energy make magnets more powerful?

You probably have more than one magnet in your home. Many people use magnets to attach notes to refrigerator doors. Some of your toys may have magnets in them. You might not realize it, but there are also magnets in things that run on electricity, such as your telephone, TV, VCR, and stereo speakers. Magnets and electrical energy are closely related.

Magnetism

What does it feel like when you try to pull a piece of iron or steel away from a magnet? The magnet exerts a force that holds on to some metal objects, and sometimes it's hard to get them loose. The ability of a magnet to exert a force is called **magnetism** (mag′ nə tiz′ əm).

We can observe magnetism at work when iron or steel objects are near a magnet. Magnets will pull on objects such as paper clips, pins, thumbtacks, or nails, because these things all have iron in them.

Powerful magnets can make work much easier for people.

Electric Current and Magnetism

When an electric current moves through a wire, it makes the wire a magnet. By itself the wire isn't a very strong magnet. You don't even notice that the wires in your home are magnetic. If a wire is coiled around and around, the magnetism of the wire becomes easier to observe. The coiled wire is a stronger magnet, but it still won't pick up much material.

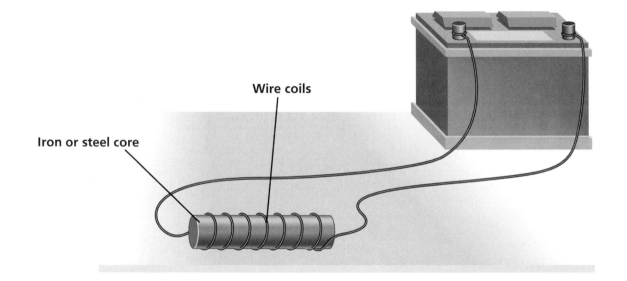

Iron or steel core

Wire coils

An electromagnet has a core of iron or steel inside wire coils. The wire coils carry electric current. The coils must be connected to a power source and form a circuit.

The wire coil becomes a much stronger magnet if you put a piece of iron inside the coils of wire. While the electric current is flowing, the iron becomes a magnet. When the current is turned off, the magnetic force stops and the objects drop from the magnet. This combination—a core of iron inside wire coils that carry electric current—is called an **electromagnet** (e lek′ trō mag′ nət).

Electromagnets are very useful because their magnetism can be turned on and off. When the electric current is flowing through the wires, the electromagnet is on. When the current is not flowing through the wires, the electromagnet is off. Large magnets like the ones that pick up cars and put them down are electromagnets. They can be turned on and off whenever the operator chooses, making it easy to pick up and put down heavy iron or steel objects.

Electromagnets can be extremely strong. Most of the strength of an electromagnet comes from the number of wire loops around the core. The more loops the wire makes, the stronger the electromagnet will be. Very large electromagnets have thousands of loops of wire coiled very closely together.

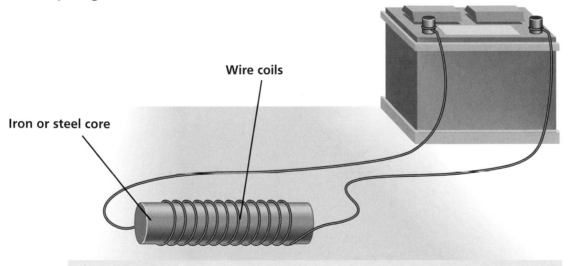

Wire coils

Iron or steel core

The strength of an electromagnet depends on the number of wire loops it has around the core. The more loops it has, the stronger its magnetism will be. Notice that this electromagnet is stronger (it is larger and has more coils) than the one on the opposite page.

How We Use Electromagnets

Electromagnets help us use electrical energy to make parts of things move. Electromagnets are part of many things we use every day. Telephones, doorbells, cars, earphones, VCRs, and cassette players all have electromagnets in them. Powerful electromagnets are used in televisions and stereo speakers.

Recycling centers use electromagnets to separate steel cans from cans that are made of aluminum. Since steel is attracted by magnetism, electromagnets help to separate

Aluminum cans aren't attracted to electromagnets. This is how recycling centers separate material they want from material they don't want.

steel cans from aluminum cans. This helps recycling centers process large amounts of material in a shorter time.

Electromagnets are also in machines that move. Many big electromagnets are used to lift and drive trains called mag-lev trains. *Mag* is short for magnetic. Mag-lev trains have no wheels. Instead, they use magnets to actually lift the train above the rails. A mag-lev train can go as fast as 500 km per hour. It moves very quickly and smoothly because it is lifted or levitated above the rail; none of it is actually touching the rail.

A mag-lev train has magnets fastened on the bottom. The strength of the magnets is adjusted so that they hold the train just above the rail. The magnets do not attach the train to the rail.

CHECKPOINT

1. What is magnetism?
2. How can electric current make a magnet?
3. How do we use electromagnets?
   How does electrical energy make magnets more powerful?

ACTIVITY

Making an Electromagnet

Find Out

Do this activity to find out how you can make your own electromagnet.

Process Skills

Observing
Communicating
Predicting

WHAT YOU NEED

1-m insulated wire with stripped ends

D-cell battery holder

D-cell battery

pencil

small paper clips

steel nail, at least 10-cm long

Activity Journal

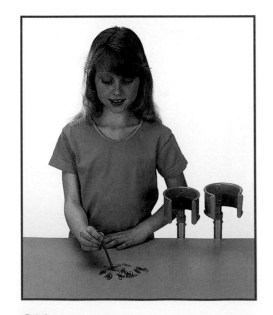

WHAT TO DO

1. Touch the nail to a pile of paper clips and see if any of them are attracted. Repeat this step with the pencil. **Record** your observations.

   Safety! *Hold the pointed end of the nail away from you.*

2. Leaving about 20 cm of extra wire at the beginning and end, wrap 15 loops of wire around the pencil. Wrap the wire around the pencil in very tight, close loops.

3. Put the battery in the battery holder. Connect one end of the wire to one end of the D-cell battery holder. Connect the other end of the wire to the other end of the battery holder. **Predict** what will happen to the paper clips. **Record** your prediction.

4. Carefully move the pencil into the pile of paper clips. Pick up as many paper clips as you can. **Observe** and **record** what happens.

5. Disconnect the wire from both ends of the battery. Repeat Steps 2 through 4, this time wrapping the wire around the steel nail instead of the pencil. **Observe** and **record** what happens.

6. Move the nail away from the pile of paper clips. Disconnect both ends of the wire from the battery holder. Count the number of paper clips you picked up and **record** the number.

CONCLUSIONS

1. What happened when you touched the pencil to the paper clips when both ends of the wire were attached to the battery?

2. What happened when you repeated the activity with a nail instead of a pencil?

ASKING NEW QUESTIONS

1. What could you change to make the nail pick up more paper clips?

SCIENTIFIC METHODS SELF CHECK

✔ Did I **predict** what would happen when I wrapped the wire around the pencil and then the nail? Did I **observe** what happened with both?

✔ Did I **record** my predictions and observations?

Static Electricity

Find Out

- What electric charges are and how they build up
- How static electrical charges behave
- The effects of static electricity

Vocabulary

electric charges
static electricity

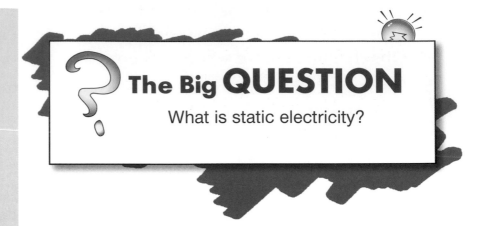

The Big QUESTION

What is static electricity?

*E*lectrical energy is invisible. You can't see it or hear it. But you can see and hear the effects of electrical energy all around you. Sometimes you can feel its effects, too. Have you ever seen clothes cling to each other when you take them out of the dryer? Have you ever heard them crackle? Have you ever gotten a shock after walking across a carpet? Then you have experienced the effects of electrical energy. This lesson explains how electrical energy builds up on objects and how it affects our lives.

Electric Charges

All matter and all living things have forces in them called **electric charges.** When objects rub together, more electric charge builds up on one of the objects than on the other.

Electric charges can make objects pull toward each other or push away from each other. You probably have seen clothing stick together or cellophane wrapping stick to a package. These effects happen because of the electric charges in the different objects. If enough electric charge builds up, a spark flies between objects. You may have seen sparks fly from a blanket when you're in bed at night.

Sometimes you can see, hear, and feel the effects of electrical energy.

Static Electricity

When electric charges build up in objects that rub together, the electricity they form is called **static** (sta′tik) **electricity.** It is called *static* because the electric charge is on something and not moving (even though it moved to get there). Something that is static stays in one place and does not move around. Static electricity doesn't flow like electric current does. Clothes that rub together when they are in the dryer have static cling when they come out. The cling is caused by static electricity.

These two balloons have different static electrical charges. The different charges pull the balloons toward each other.

The electrical charges in the balloons in this picture are alike. The like charges are pushing the balloons away from each other.

Charges Stick Together or Push Away

Static electrical charges make objects stick together or push away from each other. What happens depends on whether the charges are alike or different.

If the electrical charges of two objects are *alike,* the objects push away from each other. If the electrical charges of the objects are *different,* the objects pull toward each other.

Static electricity is a buildup of extra electrical charge in one place. How could you get rid of the extra electric charge? The easiest and safest way to get rid of it is by transferring it to the ground. Getting rid of static electricity this way is called *grounding.*

Effects of Static Electricity

The sparks and cling caused by static electricity around your home can be annoying, but they are not dangerous. However, big sparks—such as lightning—can give a very dangerous shock that can cause death or start fires. Static electricity builds up in clouds as they move through the air. When that electricity travels between the clouds and the ground, you see lightning.

Lightning will hit the tallest things in an area first. People sometimes put lightning rods on the roofs of buildings to carry the electric current away from the buildings. This works because the rod is connected to the ground with a wire.

Lightning

Even a small amount of static electricity can be dangerous to electronic machines, especially computer systems. A computer chip can be damaged by a sudden change in electric current. If a spark jumped to the chip, it could destroy the chip. People who make or repair computers must be very careful not to become electrically charged. This might happen when they walk on carpet or when clothing rubs together.

A lightning rod grounds the huge static electric charge of a lightning bolt.

CHECKPOINT

1. What are electric charges and how do they build up?

2. How do the different static electrical charges behave?

3. What are the effects of static electricity?

 What is static electricity?

ACTIVITY

Observing Static Cling

WHAT YOU NEED

ruler

transparent
tape

*Activity
Journal*

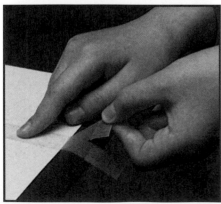

WHAT TO DO

1. Measure and tear two strips of tape about 10 cm long off the tape dispenser.

2. Stick both strips to your desk top, leaving about 1 cm over the edge. Fold the edge back so that there's a nonsticky part to grab.

3. Gently peel the strips off your desk, one at a time, so that the tape doesn't curl up.

4. Hold the two strips by their ends and bring them close to each other. **Observe** what happens. **Record** your observations.

5. Have another student stroke both of the strips several times with his or her fingers. Bring the strips together again. What happens? **Record** your observations.

6. Stick one of the strips back on your desk. Now, stick the other one right on top of it.

7. Peel both strips off your desk. Now, peel the two strips apart. Bring the two strips together again. What happens? **Record** your observations.

CONCLUSIONS

1. What happened when you brought the strips of tape near each other in Step 4? In Step 7?

2. What happened after the other student stroked the strips?

ASKING NEW QUESTIONS

1. Why do you think you got different results depending on what you did to the tape?

2. Why did stroking the strips have the effect it did?

SCIENTIFIC METHODS SELF CHECK

✔ Did I **observe** the effects of static electricity on the tape strips?

✔ Did I **record** my observations?

Review

Reviewing Vocabulary and Concepts

Write the letter of the word or phrase that completes each sentence.

1. The energy that flows through a wire to make your refrigerator work is ___.
   a. static electricity **b.** mechanical energy
   c. electrical energy **d.** a pathway

2. The pathway for an electrical current is called a ___.
   a. nonconductor **b.** circuit
   c. generator **d.** battery

3. Material through which electricity flows easily is a ___.
   a. producer **b.** conductor
   c. wire **d.** nonconductor

4. Material through which electrical energy does not flow well is a ___.
   a. nonconductor **b.** current
   c. circuit **d.** wire

Match each definition with the correct term.

5. the ability of some metal to attract other metal **a.** magnetism

6. extra coils around the core of this make it stronger **b.** wire loops

7. create a magnet when combined with an iron core **c.** electromagnet

8. what makes socks stick together when they come out of the dryer **d.** static electricity

Understanding What You Learned

1. List three kinds of activities you enjoy that use electrical energy.

2. Where is electric power in a streetlight changed into light?

3. What kind of energy comes out of a flashlight? An electric blanket?

4. What can add strength to an electromagnet?

5. What moves from one balloon to another to make them stick together?

Applying What You Learned

1. How are conductors and nonconductors helpful in getting power to your classroom?

2. How can we protect ourselves from being hurt by electricity?

3. Explain how you could get the power of wind turning windmills to run machines.

4. What is the difference between static electricity and the electricity that makes a toaster work?

 5. What would be the best way to show how much we depend on electrical energy?

For Your Portfolio

Imagine that you have built a tree house next to your house. You want to be able to spend the night in it and read yourself to sleep. Explain how you would bring electricity, with the help of an adult, to the tree house.

Unit Review

Concept Review

1. Explain three changes that can happen when two materials are combined.

2. Describe the different ways that people, plants, and animals use light.

3. Explain how two simple machines make work easier to do.

4. How is static electricity different from other electricity?

Problem Solving

1. How can scientists use their knowledge of chemical changes to make new products?

2. Imagine that you are an inventor. What invention can you think of that could help people use the sun's energy?

3. How many simple machines do you have in your school or neighborhood? Find and list at least eight examples of levers, inclined planes, or other simple machines.

4. How could you find out more about safety during electrical storms?

Something to Do

Cut out pictures from magazines and newspapers that illustrate as many of these concepts as you can find.

- matter changing from liquid to solid
- matter changing from liquid to gas
- a pulley
- an inclined plane
- a physical change
- a chemical change
- fuel a generator can use to make electricity
- electrical energy being transformed into light energy
- a reflection
- a prism
- electrical energy being transformed into thermal (heat) energy
- electrical energy being transformed into mechanical energy

Make a collage of the pictures and label them with the concepts.

UNIT D

Health Science

Muscles and Bones Work Together

What kinds of sports activities do you enjoy? Do you like to skate, ride a bike, or swim? Do you and your friends play baseball, basketball, or soccer? To play many kinds of sports, you must work as part of a team. Did you know that the parts of your body work together as a team too? To play sports, eat, sleep, or do anything else, you use many different parts of your body. You can think of the different systems of your body as team players that work together to keep you alive and healthy. In this chapter, you will learn about two of your body's most important team players, your muscles and your bones.

The Big IDEA

Muscles and bones support our bodies and allow us to move.

CHAPTER SCIENCE INVESTIGATION

Discover what muscles and bones you use when you exercise. Find out how in your *Activity Journal.*

Muscles and Bones

Find Out

- What muscles do
- What bones do
- What holds muscles and bones together

Vocabulary

muscular system
calcium
skeletal system
skull
ligaments
tendons

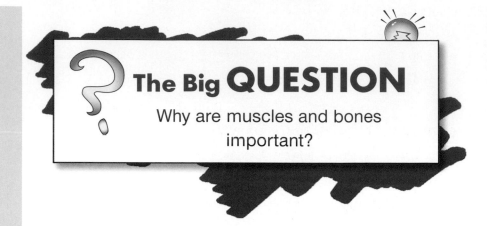

The Big QUESTION

Why are muscles and bones important?

Open one hand and wiggle your fingers. Now make a fist. Whenever you move your hand, two important team players are at work—your muscles and your bones. Let's find out how they work.

Your Mighty Muscles

If you shut your eyes and feel one of your hands, you will feel that it has a lot of bones in it. Bones in your hand stay in place and work for you because of muscles. Muscles are like thick rubber bands. They help hold your hand bones together and allow your hand to move.

Muscles also help give shape to your body. You can feel the large muscles in your arms and legs. These are especially important when you run and play. The more you run, play, and use your muscles in other ways, the stronger and larger they get.

Muscles are tough, strong fibers that lengthen and shorten to help move your body. Have you ever tried to do a back bend or the splits? If you have, you know that you can bend only so far. You cannot stretch farther than your muscles will let you go.

You have muscles all over your body—more than 600 all together. All of your muscles make up your **muscular system** (mus kyə⁄ lər sis⁄ təm). Your muscular system has exactly the same number of muscles now as when you were born. Of course, as you get bigger, your muscles get bigger too. You have many different muscles that work together to help you move around.

Muscles are always growing and changing. Muscles must be used to keep them strong. If the muscles in the body are not used enough, they will shrink in size and lose strength.

Muscles make up almost half of your body's weight. Like other parts of your body, muscles work best when you eat good food, get enough rest at night, and exercise, exercise, exercise! Your muscles are just waiting for you to do the right things so that they can grow bigger and stronger.

The Bones, Muscles, and Tendons of the Hand

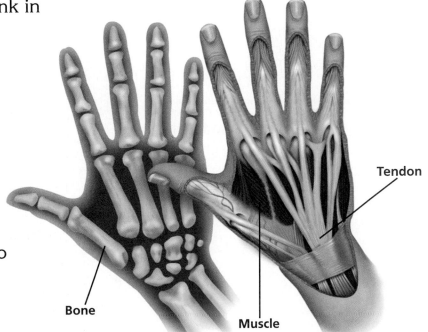

Tendon

Bone

Muscle

Bones Hold You Up

Without bones, you would be as limp as a rag doll. Bones support your body and allow you to do important things, such as reach for a book on a high shelf or swing a bat. Your bones began to grow even before you were born. Like every other part of your body, bones grow and change. They are one of your body's fastest growing parts. Bones not only grow larger, they are always replacing old parts of bone with new growth. They will continue growing and changing as long as you live.

Bones help your whole body by storing calcium. **Calcium** (kal′ sē əm) is a mineral your body needs to stay healthy. Your body stores the most calcium during the years when your bones are growing. Growing children need calcium-rich foods like milk, cheese, and beans in their diets. Getting enough calcium in your diet will help keep your teeth and bones healthy and strong.

As you grow taller and taller, your bones help give shape to your body.

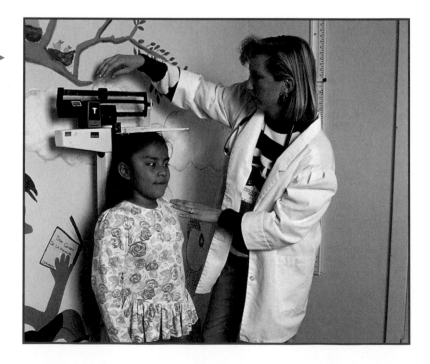

You need muscles and bones to walk and run.

All your bones together make up your **skeletal system.** The long bones in your legs and the bones in your back help make you as tall as you are. You have many short bones, too. There are short bones in your hands and feet. They allow you to walk, jump, tie your shoes, and do many other things.

The bones in your head make up your **skull.** The skull protects a very important part of your body—your brain. Your ribs are bones that protect your heart, lungs, and other body parts.

Like muscles, bones grow best when you take good care of your body. Eating lots of good food, especially milk and dairy products that have calcium, is important. Bones grow best when the body is active. Any physical activity, such as running, jumping, and dancing, will help bones grow strong.

Teamwork

On a soccer team, midfielders and forwards work together to score goals. Like players on a soccer team, bones and muscles team up to move you around and let you do the things you do. They are helped by two other sets of players, the ligaments and the tendons.

Ligaments

Ligaments (lig′ ə mənts) are strong tissues that connect parts of the skeleton together, bone to bone. Bones that meet at joints, like the knees and elbows, have more ligaments than bones that meet at other places in your body. If you feel a pain in your knee or elbow, it may be the ligaments that are hurting.

It takes teamwork to score points.

Tendons

Tendons are strong cords of tissue that help muscles move bones. One end of a tendon is attached to a muscle and the other end to a bone. So the words of the song that say "The hip bone's connected to the thighbone" are not quite right. Like the elbow, the place where the hip and thighbones meet is a joint. A web of ligaments connects the bones. To move a leg, a tendon attached to both the leg bone and the leg muscles must do its work.

The muscles and the bones must work as a team to move you around. The muscle players have many different names. So do the bone players. Tendons and ligaments also help. Working together, they allow you to dance, run, swim, or do whatever else you want to do.

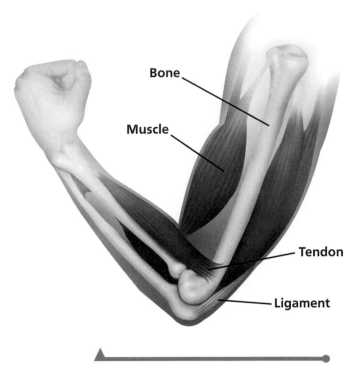

Bone

Muscle

Tendon

Ligament

Muscles, bones, tendons, and ligaments help you move.

CHECKPOINT

1. What do muscles do?

2. What do bones do?

3. What holds muscles and bones together?

 Why are muscles and bones important?

ACTIVITY
Making Bones Steady

WHAT YOU NEED

two pieces of string about 50 cm each

flexible drinking straw

tape

Activity Journal

WHAT TO DO

1. Working with a partner, tie one end of each piece of string to one end of the straw.

2. Tape the other end of the straw to the table as shown.

3. One partner should hold the loose end of one string while the other holds the other string. You both should hold the string by keeping your hands flat on the table, pressing down on the string.

4. **Predict** what you will have to do to each string to keep the straw steady.

5. Using only the strings, work with your partner to steady the straw so it stands up straight.

6. **Observe** what happens as you and your partner pull on the strings. **Record** your observations.

CONCLUSIONS

1. Compare your prediction with your observations.
2. How do the strings pulling on the straw compare to muscles pulling on a bone?

ASKING NEW QUESTIONS

1. What would happen if you and your partner competed instead of working together?
2. What advice would you give someone trying this activity for the first time?

SCIENTIFIC METHODS SELF CHECK

✔ Did I **predict** what would happen to the straw?

✔ Did I **observe** what happened?

✔ Did I **record** my observations?

How Muscles Move Bones

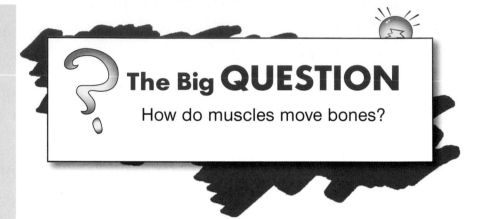

The Big QUESTION

How do muscles move bones?

If you could look inside your body, you would be amazed. You would see that your skeleton is covered with muscles. Though the muscles come in all sizes and shapes, they all do one thing. What do you think that might be?

Muscles at Work

As a muscle pulls on bone, the muscle contracts. It gets shorter and looks fatter. Make a fist and squeeze it. Now bend your elbow and raise your arm. A muscle will pop up on the top of your arm. The muscle is pulling on a bone in your arm. It bulges because it is **contracting**—pulling together and getting shorter. Some people have bigger muscles than others, especially people who lift weights. Even though your muscles may not appear to be large, they do what all muscles do. They contract as they work for you.

Many muscles do not work alone. They work in pairs. One muscle pulls a bone toward itself. Its partner relaxes, allowing the bone to be pulled away from it. Many different muscles must work together to move your body.

You have two types of muscles working for you. One type of muscle moves bones. The other type of muscle works for you in a different way.

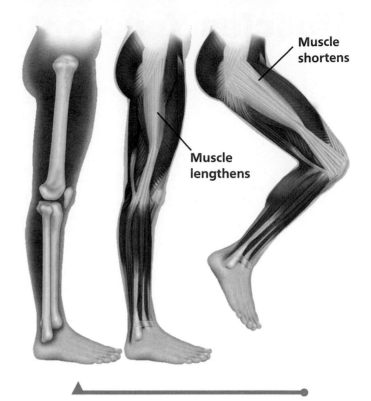

Muscle shortens

Muscle lengthens

Muscles and bones work together to help you move.

Lifting weights helps build muscles.

Muscles You Control

Voluntary muscles are the muscles you control. They help move your skeleton around. Your brain sends a message to the nerve endings in the muscle. The muscle does what your brain tells it to do.

Here's how it works. Imagine you want to scratch your head. Your brain sends a message. The message goes to the nerve endings in the muscle in the top of your arm. The muscle contracts, or shortens. This shortening of the muscle moves the bones of your arm. Your whole arm moves, bending at the elbow. Now you can scratch your head.

Your brain tells your voluntary muscles what to do, and they do it.

If you blow a trumpet, your voluntary muscles are working for you.

Suppose you have finished scratching your head. You decide to pick up your pencil. It seems to happen instantly. However, first a muscle behind your elbow must shorten or contract. Another muscle, its partner, relaxes and gets longer. Working together, these muscles straighten your arm. Now you can use the muscles in your hand to pick up the pencil.

Voluntary muscles make it possible for you to do all sorts of movements. Skating, walking, jumping, and bending are all voluntary movements. Even turning over in your sleep is a voluntary movement!

Muscles That Work on Their Own

Some muscles work without you thinking about them. These muscles are called **involuntary muscles.** The heart is an involuntary muscle. It pumps blood to your body by receiving automatic signals from the brain. You do not have to think to make your heart beat. Your breathing is also controlled by involuntary muscles. As the muscles around your lungs contract and relax, you breathe without having to think about breathing.

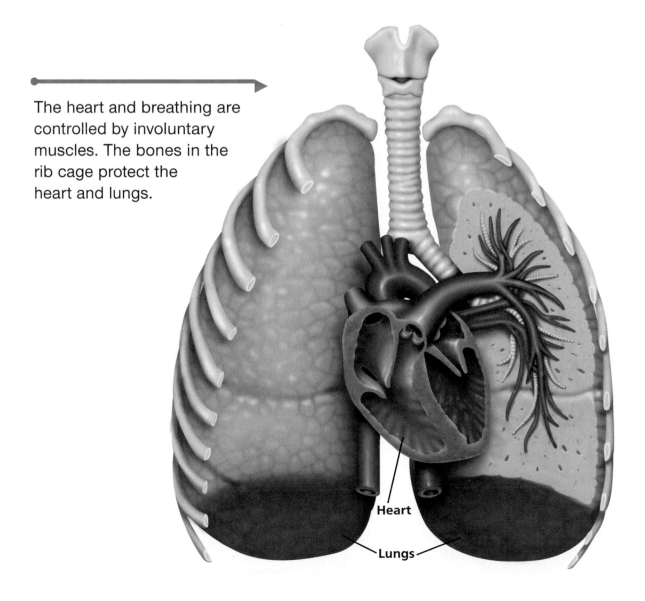

The heart and breathing are controlled by involuntary muscles. The bones in the rib cage protect the heart and lungs.

Heart

Lungs

The muscles that push food down the tube to your stomach are also involuntary muscles. So are the muscles that push food through your digestive system. Thank goodness for these involuntary muscles! We would never get any rest if every second we had to tell our heart, our digestive system, and our other involuntary muscle systems what to do.

Involuntary muscles usually are not connected to bones. However, they are protected by bones. For example, the heart and lungs are protected by the rib cage.

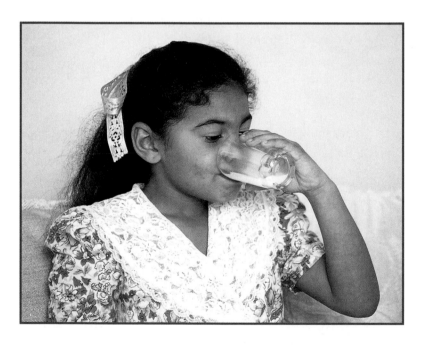

Voluntary muscles allow you to lift a glass and swallow. Involuntary muscles push foods through your digestive system.

CHECKPOINT

1. How do all muscles work?
2. What do voluntary muscles do?
3. What are involuntary muscles?

 How do muscles move bones?

ACTIVITY

Making Muscles and Bones Work Together

Find Out

Do this activity to learn what muscles and bones must do before you move.

Process Skills

Hypothesizing
Predicting
Observing
Communicating

WHAT YOU NEED

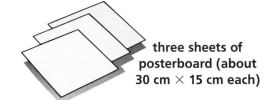
three sheets of posterboard (about 30 cm × 15 cm each)

chenille wire

black marker

two long balloons

Activity Journal

hole punch

masking tape

Activity Journal

WHAT TO DO

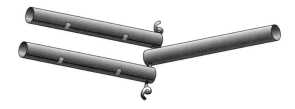

1. Work with a partner. Roll each piece of posterboard into a tube. Tape them so that they will not unroll.

2. Punch two holes in opposite sides of one end of each tube.

3. Lay the tubes down. Line up the holes. Thread the wire through, and twist the ends so the wire stays in place.

4. Tape the ends of the top and bottom tubes together as shown.

5. With the marker, write "A" on one balloon and "B" on the other.

6. Blow up the balloons part way and tape them to either side of the end of the center tube as shown. Tape the free ends of the balloons to the other tubes as shown.

7. **Predict** what will happen to the balloons when you try to straighten out the tubes.

8. **Observe** what happens as you and your partner straighten out the tubes. Try moving the tubes in different ways.

9. **Record** your observations.

CONCLUSIONS

1. Compare your predictions with your observations.
2. What part of the body is each balloon like?
3. What part of the body are the tubes like?

ASKING NEW QUESTIONS

1. How are the tubes and balloons like bones and muscles?
2. Is it possible to move the tubes without moving the balloons? Why or why not?

SCIENTIFIC METHODS SELF CHECK

✔ Did I **make a hypothesis** when I **predicted** what would happen to the balloons?

✔ Did I **observe** what happened to the balloons?

✔ Did I **record** my **observations?**

The Major Bones and Muscles

Find Out

- What bones are like on the outside and on the inside
- What some different kinds of bones are
- What some different kinds of muscles are

Vocabulary

cardiac muscle
skeletal muscles
smooth muscle

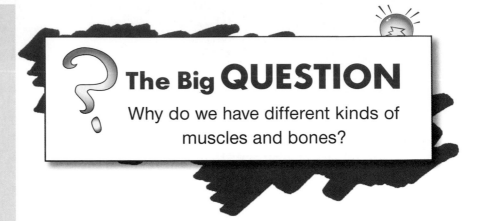

The Big QUESTION

Why do we have different kinds of muscles and bones?

You know that your muscles and bones are a great team. It takes different kinds of muscles and bones to move you around. What might some of them be?

What Is in a Bone?

Bones are hard and strong. But have you ever picked one up? It feels very light. That's because most bones have hollow places inside them. If they were solid all the way through, they would weigh much more—so much more, in fact, that you would not be able to move your own skeleton!

Most of your skeleton is made of hard material. Minerals such as calcium make the hard part of a bone strong. Hard bone is full of tiny openings. Blood vessels travel through the openings, bringing oxygen and nourishment to your bones. The inside of a bone is soft and full of holes, like a sponge. The spongy part of your bones makes red blood cells for your body.

Your bones are alive. As you grow, they do too. Bones in your arms and legs get longer until you reach full size. The bones of your skeleton will stop growing longer when you are a full-grown adult, but they will keep repairing themselves all your life.

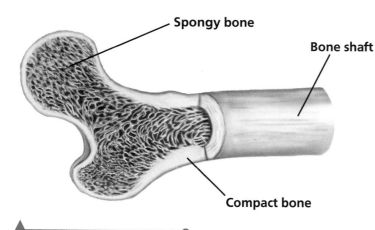

The inside of a bone is full of tiny openings.

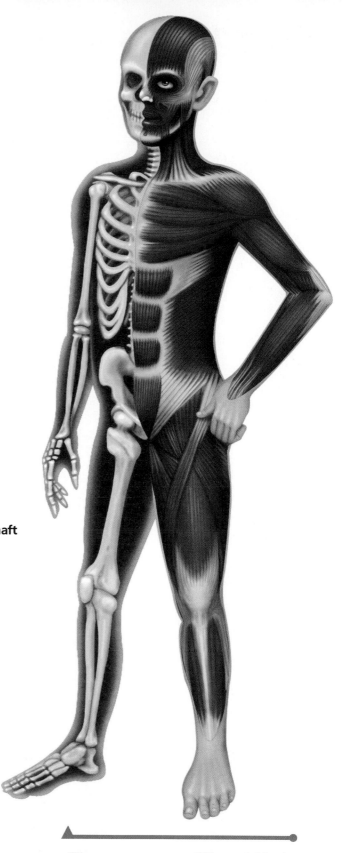

There are many different kinds of muscles and bones inside the body.

Four Kinds of Bones

You have four kinds of bones. Think about what your skeleton looks like. Where would the longest bones be? As you might guess, the bones in your arms and legs are the longest. They are called long bones. Picture the bones in your hands and feet. They are shorter than those in your arms and legs. They are called short bones.

Every bone has a function. Together they help you throw a ball, run, eat, wiggle your fingers and toes, and do hundreds of other things.

Can you tell which bones are arm and leg bones and which are bones from the hand and foot?

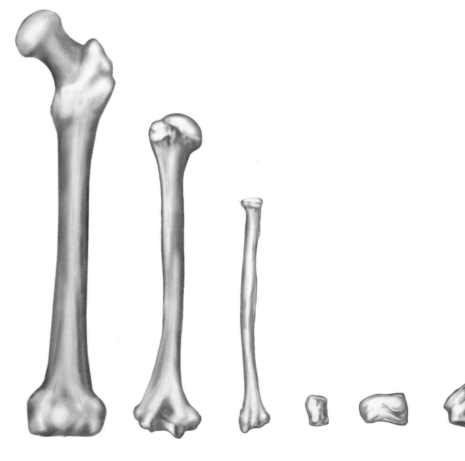

Long bones　　　　**Short bones**

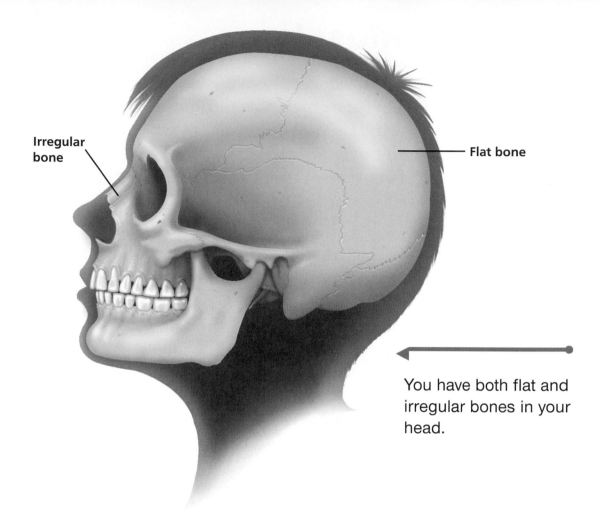

Irregular bone

Flat bone

You have both flat and irregular bones in your head.

You have irregular bones in your backbone that move as you move your body. You also have irregular bones in your face and in your middle ear. The bones in your face do not move at all. The hollow places in your face bones protect your eyes. The bones that move in the middle ear change sound waves into sound you can hear.

Your ribs are flat bones. So are the bones in your skull. The rib bones protect your heart and lungs. The skull bones protect your brain. Flat bones do not move. The only part of the skull that moves is the lower jawbone, which is not a flat bone.

Three Kinds of Muscles

Muscles help move your body. They also help move materials through your body. Your heart is a very important muscle. As it contracts, it pumps blood to all your body parts. The muscle that makes up the heart is called **cardiac** (kär′ dē ak′) **muscle.** Cardiac muscle is involuntary muscle.

The muscles that move the bones in your skeleton are called **skeletal muscles.** You can make your skeletal muscles move. All you have to do is think and do it. Skeletal muscles move at your command.

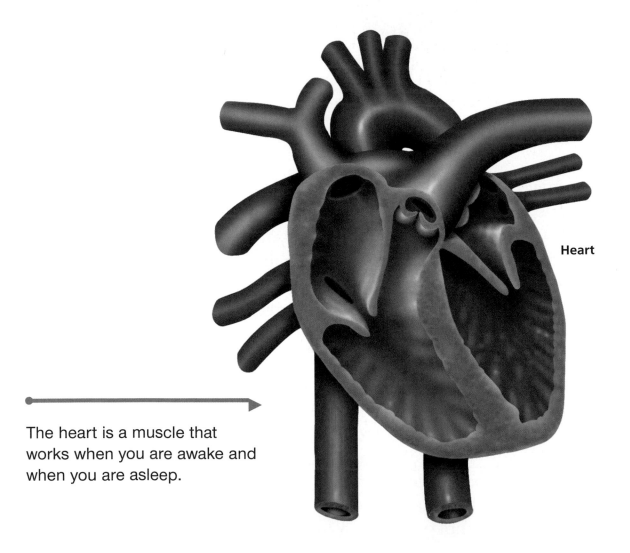

Heart

The heart is a muscle that works when you are awake and when you are asleep.

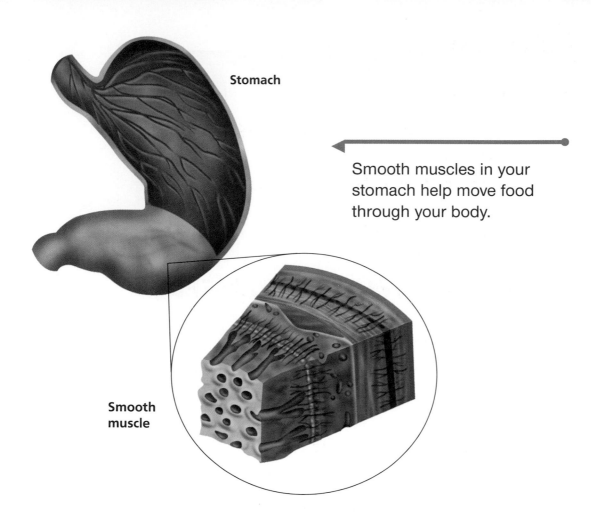

Stomach

Smooth muscles in your stomach help move food through your body.

Smooth muscle

A third type of muscle is **smooth muscle.** Smooth muscles move food through your body. Smooth muscles are also involuntary muscles that work whether you think about them or not.

CHECKPOINT

1. What are bones made of?

2. What are four different kinds of bones?

3. What are three kinds of muscles?

 Why do we have different kinds of muscles and bones?

ACTIVITY

Which Muscles Will Work?

Find Out

Do this activity to learn what happens when your mind tells your body to do something.

Process Skills

Predicting
Observing
Communicating

WHAT YOU NEED

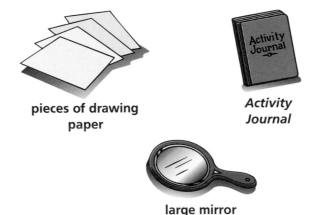

pieces of drawing paper

Activity Journal

large mirror

WHAT TO DO

1. Work with a partner. You will take turns giving your partner directions. You will draw a picture before you follow each direction.

2. Sit facing your partner. One of you will hold the mirror and read a direction. The other partner will fold hands in lap and listen. The directions are:
   - Tug your ear.
   - Slap your knee.
   - Touch your opposite shoulder.
   - Flap one arm like a wing.

3. **Predict** which muscles will have to work to make each motion. **Draw** a picture of yourself, with an arrow pointing to the muscles you think will work when you follow the direction.

4. Ask your partner to read the directions again, one at a time. Make each motion after you hear each direction. **Observe** yourself in the mirror to see and then feel if your prediction was right.

5. **Record** your observations.

Conclusions

1. What helped you predict which muscles would work?

2. How do you think you might become better at predicting which muscles will work?

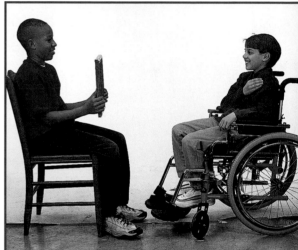

Asking New Questions

1. How many muscles all together do you think were needed when you followed each direction?

2. How might you find out the names of the muscles you used?

SCIENTIFIC METHODS SELF CHECK

✔ Did I **predict** which muscle would work?

✔ Did I **observe** what happened to the muscle?

✔ Did I **record** my observations?

Review

Reviewing Vocabulary and Concepts

Write the letter of the word or phrase that completes each sentence.

1. The framework of bones that supports your body is called the ___.
   a. skeletal system   **b.** joint
   c. skeletal muscle   **d.** muscular system

2. Muscles are attached to bones by tough, elastic tissues called ___.
   a. joints   **b.** involuntary muscles
   c. tendons   **d.** cardiac muscle

3. Muscles that you control are ___.
   a. cardiac muscle   **b.** joints
   c. tendons   **d.** voluntary muscles

4. The muscles that move food through your digestive system are called ___.
   a. smooth muscles   **b.** your muscular system
   c. voluntary muscles   **d.** cardiac muscle

Match the definition on the left with the correct term.

5. a mineral that helps keep your bones strong

6. muscles that work on their own, without any direction from you

7. the type of muscle that moves your bones

8. the heart muscle

   a. involuntary muscles

   b. calcium

   c. cardiac muscle

   d. skeletal muscle

Understanding What You Learned

Write the word or phrase that completes each sentence.

1. Your muscles and bones work together to support your body and allow it to ___.

2. An area where two bones meet, allowing your body to bend, is called a ___.

3. To pull a bone one way and then another, muscles must work ___.

Applying What You Learned

1. What does the skull protect?

2. Name three bodily activities that use involuntary muscles.

3. What can you do to make your muscles strong?

 4. Why are muscles and bones important to our bodies?

For Your **Portfolio**

Try to keep track of everything you do in a short amount of time—for example, during lunch at school or for the first half hour when you come home. In a notebook, enter the date and time of day, and write a description of each activity you performed using voluntary muscles. Don't forget even the smallest activities, such as scratching an itch or moving your toes. Do this again for another time of the day. Then compare your diary entries. Do you use muscles in different parts of your body at different times of the day?

Personal and Community Health

Have you ever heard someone say "Look before you leap"? This saying teaches a message about safety. Suppose you wanted to jump over a puddle. You should first see what lies on the other side. Maybe there is a stone or broken glass. Will you land on it if you jump across? "Look before you leap" is a message that means "be careful." It also means plan ahead.

People plan ahead when they wear safety equipment. They plan for accidents that might happen. Safety equipment gives protection to your body. People wear safety equipment when they work and play.

The Big IDEA

An important part of being healthy is practicing safe habits.

CHAPTER SCIENCE INVESTIGATION

Investigate the causes of common injuries. Find out how in your *Activity Journal.*

Avoiding Injuries

Find Out

- How a bicycle helmet helps keep you safe
- How sports equipment helps keep you safe
- How work equipment helps keep you safe
- How seat belts help keep you safe

Vocabulary

injuries

helmet

seat belts

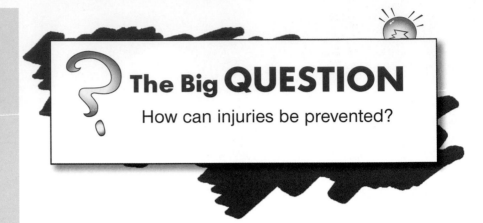

The Big QUESTION

How can injuries be prevented?

When you play, you try to be safe. But sometimes accidents happen. People get scratched, cut, and bruised all the time. You can learn habits that will help prevent accidents and keep you safe.

Bicycle Helmets

Riding a bicycle is good exercise. But sometimes, bicycling can be dangerous. Cuts, scrapes, and other **injuries** can happen while you are riding your bicycle.

You can ride your bicycle more safely if you wear a sturdy **helmet** on your head. A cyclist can fall and hit her or his head on the ground or pavement. Wearing a helmet can help prevent a serious head injury. The hard plastic shell spreads the blow over the whole helmet and softens the blow to your head.

The shell prevents sharp objects from passing through it and keeps the helmet from breaking.

You need to be sure that your helmet fits you right. It should fit snugly. You should not be able to stick a finger between your head and the helmet.

The straps in front of your ears and those in back should form a tight V when the chin buckle is fastened. You should just be able to fit your thumb between your chin and the chin strap. If you can jiggle the helmet, it is too large for you. To test for protection, put your helmet on and then slowly move your face toward a wall. The front of the helmet should touch the wall before your nose does.

A helmet is made to protect your head against only one big blow. You should get a new helmet after a serious accident. If a helmet becomes cracked or damaged, do not wear it or allow anyone else to use it. A damaged helmet cannot protect your head.

A helmet must fit straight on your head and low on your forehead to protect the front of your head.

Safe Sports Equipment

Sports players wear helmets and other safety equipment. When you play softball, you should wear a batting helmet with a face guard. The guard helps protect your eyes and nose. Some players, like catchers and goalies, should wear even more safety equipment for extra protection.

A football player's helmet should fit snugly without the chin strap fastened. The strap should snap to the helmet at four places. The back of the helmet should have enough padding to keep the hard shell from touching the neck. A football player should wear shoulder, hip, spine, and thigh pads and a mouth guard to protect the teeth.

A face guard helps protect the eyes and nose.

Wrist guards protect the wrists during a fall.

Skaters who fall can injure their heads. They can also scrape their knees. Wrist guards help support skaters' wrists to keep them from breaking during a fall. In-line skaters and skateboarders wear helmets, elbow pads, knee pads, and wrist guards.

Shin guards help protect legs from being injured.

Soccer players should wear shin guards. Injuries can happen when players miss the ball and accidentally kick each other. A strong kick could break a bone in the lower leg. Shin guards help prevent leg injuries.

Swimmers should wear goggles to protect their eyes. Goggles prevent burns from the chemicals used in swimming pools. These burns make your eyes feel dry and scratchy. Goggles should be tight enough to keep out water.

Safe Work Equipment

Many workers wear safety equipment. Construction workers build tall buildings. Sometimes pieces fall from the floors above. To be safe, construction workers wear stiff plastic or metal helmets called hard hats. These helmets protect their heads. Some workers heat metal parts together. They wear safety goggles. The goggles keep chips and sparks out of their eyes.

Sometimes people use safety equipment at home. People who mow lawns should wear safety goggles and ear plugs. Lawn mowers can hit objects and make them fly into the air. Sticks and stones can fly into the face of someone nearby. The loud noise of the lawn mower can damage hearing. It is important to be very careful around lawn equipment.

Even gloves are important for safety. You may wear gloves to keep your hands warm in cold weather. Cooks wear oven mitts to protect their hands from burns. Carpenters often wear gloves to prevent cuts and splinters.

Safety goggles protect your eyes from flying objects.

Seat Belt Safety

People driving and riding in cars need safe habits. Car accidents happen often. Cars are heavy and fast moving. When car drivers push the brake pedal very suddenly, the car stops, but passengers' bodies keep moving forward. Their heads could hit a window. They could hit the seat in front of them. **Seat belts** protect passengers by holding them in place. People riding in cars should wear seat belts at all times. Babies and small children should always ride in car seats. In a car crash, an air bag opens up. The air bag is like a cushion. The passenger's forward motion is stopped by the air bag.

Keep your seat belts buckled while riding in a car. Small children should be in car seats.

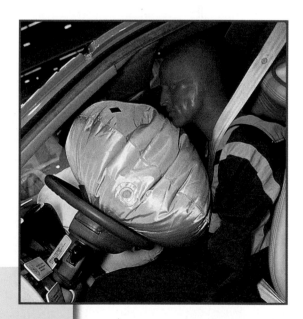

In a car crash, an air bag opens up.

CHECKPOINT

1. How does a bicycle helmet help keep you safe?

2. Explain how three different kinds of sports equipment help keep you safe.

3. Explain how three different kinds of work equipment help keep workers safe.

4. How does a seat belt help keep you safe?

 How can injuries be prevented?

ACTIVITY
Modeling Air Bags

Find Out

Do this activity to see how air bags protect you when a car you are riding in comes to a sudden stop.

Process Skills

Predicting
Constructing Models
Observing
Communicating

WHAT YOU NEED

four to six hard-boiled eggs

safety goggles

Activity Journal

twin bed sheet

WHAT TO DO

1. Put on your safety goggles. With three or four other children, stand and hold onto the short sides of a sheet. Stretch the sheet out and hold it tightly in place.

2. **Predict** what will happen when an egg is gently tossed onto the center of the sheet. **Record** your prediction.

3. **Make a model** of an air bag by *gently* tossing an egg on the center of the sheet. Take turns tossing the egg and holding the sheet. **Observe** what happens each time an egg is tossed. **Record** your observations.

CONCLUSIONS

1. Does the egg break when it lands on the sheet? Why or why not?
2. Would it break if it fell on the floor? Why or why not?

ASKING NEW QUESTIONS

1. What happens when the sheet is pulled tighter?
2. What happens when the sheet is held loosely?
3. What happens when the egg is thrown harder?
4. How is the sheet like an air bag?
5. What does the egg represent?

SCIENTIFIC METHODS SELF CHECK

✔ Did I **predict** what would happen when I tossed the egg?

✔ Did I **observe** what happened? Were my predictions correct?

✔ Did my group **make a model** of an air bag with a sheet?

✔ Did I **record** my observations of what happened?

Basic First Aid

Find Out

- What first aid is
- How to give first aid for cuts
- How lo give first aid for sprains
- How to give first aid for nosebleeds

Vocabulary

first aid

sprain

nosebleed

The Big QUESTION

What are basic first aid procedures?

Suppose you are running on a playground during recess. Your shoelaces come untied and you trip. You fall and cut your knee. Your injury is not serious, but it hurts. What should you do next? In this lesson, you will learn how to treat small injuries.

First Aid

Many injuries can be treated right away. If you get an injury, you should tell a parent, teacher, school nurse, or another responsible adult. They can treat many injuries that are not too serious.

First aid is the care given right away for an injury. Your skin can heal itself, but sometimes it needs help. Cuts and skinned knees are injuries that may need first aid. A sprain may need first aid. First aid can help heal an injury.

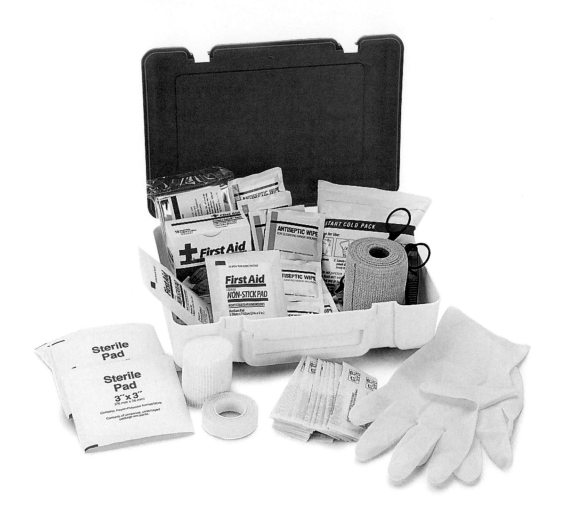

Your home or school probably has a first aid kit handy. The kit contains things that can be used to treat small injuries. It might have different sizes of bandages. Some could be used to cover wounds or to stop bleeding. Others could be used to support injured body parts or to control swelling.

The kit might have pads soaked in first aid chemicals. These chemicals clean dirt and germs from the wound. The kit might also have a first aid cream. When the cream is put on a wound, it helps keep germs from getting into the wound. Keeping cuts clean and protected will help prevent infection.

All of the things in a first aid kit can be used to treat small injuries.

Cuts

If you get cut, first you should tell a responsible adult. Never touch a cut or any blood from another person. Germs can get in the cut, so it should be gently washed with soap and water. Then it should be patted dry with a clean pad or cloth. A first aid cream can be put on the cut.

A bandage should be put over cuts that are sore or bleeding. The bandage will keep the cream in place and will help protect the area. It will soak up blood from the cut. You need to change the bandage every day. When a scab forms, you can take off the bandage. The air will help your skin heal.

Press down on a cut to slow the bleeding.

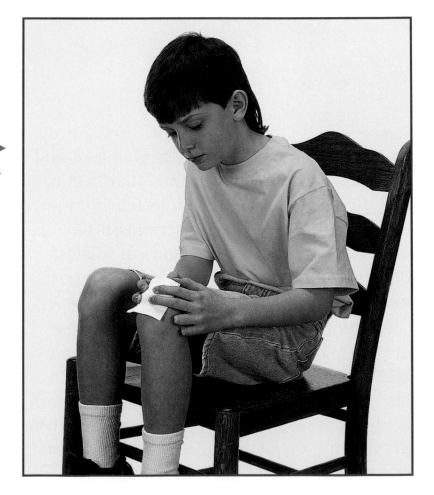

Hold a cut higher than the level of your heart.

Suppose you have a deep cut. While you are waiting for help, you can start to give yourself first aid. Put a clean cloth over the cut. Press down on it. The pressure will help slow down the bleeding.

You can slow the bleeding of a cut in another way, too. A body part that is cut should be elevated, or held up, higher than the level of your heart. If the cut is in your leg, you should lie down. Lift up your leg and rest it on a pillow or chair. If the cut is in your arm, hold your arm up and rest it on something higher than your heart.

Sprains

A **sprain** is an injury to a ligament, near a joint. Sprains often happen to an ankle, wrist, finger, or knee joint. A sprain makes your joint swell up and hurt when you move it or put weight on it. It should be elevated. An ice pack should be placed on the sprain as soon as possible. Do not try to walk with an ankle or knee sprain. Some sprains may need professional medical attention.

A sprain needs rest to heal. Keep a sprained joint still and elevated as much as possible after an accident. Apply ice to the area often for the first two or three days to help keep swelling down. After two or three days (when there is no more swelling), heat can be applied to the sprain to help the injury heal.

Elevate a sprained joint. Put a cold pack on the joint to keep the swelling down.

Nosebleeds

A **nosebleed** can happen for different reasons. If your nose gets hit, it might bleed. Blowing your nose often when you have a cold may cause a nosebleed. Walking, talking, laughing, and blowing the nose can make a nosebleed worse.

To stop a nosebleed, sit down, lean slightly forward, and pinch your nostrils shut for five to ten minutes until the bleeding stops.

If the bleeding continues, place cold cloths on your nose. Wash your hands with soap and water after the bleeding stops. If the bleeding does not stop after 30 minutes, you may need professional medical attention.

To stop a nosebleed, lean forward and pinch your nostrils.

CHECKPOINT

1. What is first aid?
2. Describe first aid for a cut.
3. Describe first aid for a sprain.
4. Describe first aid for a nosebleed.

 What are basic first aid procedures?

ACTIVITY

Making a First Aid Kit

Find Out

Do this activity to learn how to make your own first aid kit.

Process Skills

Communicating

Classifying

WHAT YOU NEED

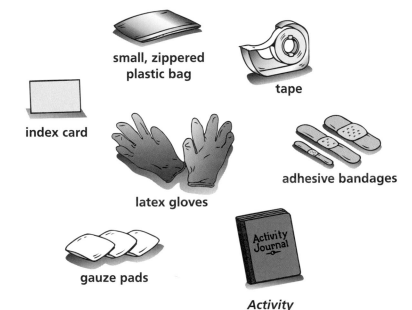

small, zippered plastic bag

tape

index card

latex gloves

adhesive bandages

gauze pads

Activity Journal

WHAT TO DO

1. Open the plastic bag.

2. List the things that can be used to treat small injuries on a card. Title the list "First Aid Kit."

3. Place the list inside the bag so that you can see it from the outside.

4. Put the things you listed inside the plastic bag. Be sure to put them behind your list.

Conclusions

1. **Explain** what each item is for.
2. What else could you put in your kit?
3. Why is a zippered, plastic bag a good thing to use for making a first aid kit?

Asking New Questions

1. Name places where you might find a first aid kit.
2. What types of workers might need a first aid kit nearby?
3. How is packing a first aid kit like packing a suitcase?
4. Where are some good places for people to keep first aid kits?

SCIENTIFIC METHODS SELF CHECK

✔ Did I **communicate** by listing the contents of the kit on the index card?

✔ Can I **identify** each item in the kit by both its name and its use?

Emergency Care

Find Out

- What rules to follow during an emergency
- What rules to follow when making an emergency telephone call
- How people are treated in an emergency room

Vocabulary

emergency
cast
stitches

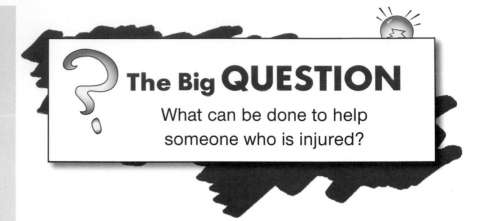

The Big QUESTION

What can be done to help someone who is injured?

Suppose you are playing on the school playground. Your friend falls off a swing. She is badly hurt. A bandage would not fix her injury. Your first aid kit is not enough help. How can you help your friend?

What to Do in an Emergency

Sometimes people are badly hurt. They need help right away. They may need more than first aid. They may need emergency care. An **emergency** is a serious event that comes without warning and calls for fast action.

You need to remember these rules to help another person who is injured.

- Stay calm during an emergency.
- Think clearly.
- Tell the injured person to lie still. Do not move the person. Moving an injured person might make the injury worse.
- Cover the person with a blanket, jacket, or sweater to keep him or her warm.
- Tell the person you will get help. Then go tell a teacher, the school nurse, or another responsible adult. They will know what to do to help.

Keep an injured person warm during an emergency.

Emergency Number

Suppose you are at home and someone falls. The person is seriously injured. Tell a parent or other adult right away. The other adult might be a neighbor or a friend's parent. Suppose your parent has an accident. If there is no other adult around, you may have to call for help. You may have to use the telephone to call 911.

Dial 911 for help during an emergency.

A 911 operator knows what kind of help is needed during an emergency.

There is a special emergency number you can dial. People who answer this number know how to help.

- Dial 911 for help.
- When the operator answers the telephone, say "This is an emergency."
- Speak slowly and clearly. Tell the operator your name. Also tell him or her what has happened and where you are. Answer all of the operator's questions.
- Listen carefully to any directions that the operator gives you. Do not hang up the telephone until you are told to do so. The 911 operator will send the help that you need.

Emergency Rooms

Sometimes an ambulance rushes an injured person to an emergency room. The person might have to be given medicine or oxygen immediately. The person's heart might need to be started again.

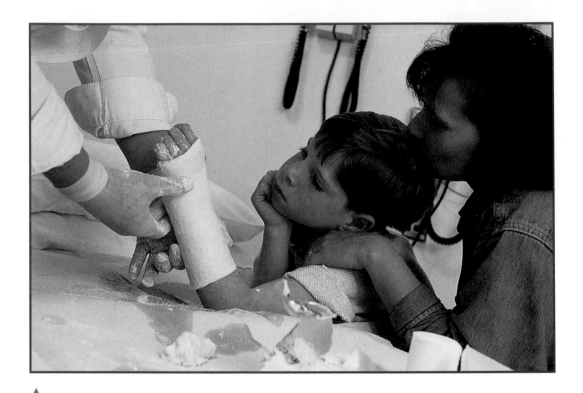

A cast helps broken bones to heal in the right way.

An emergency room is a very busy place. Emergency-room workers handle many different kinds of injuries. For example, a bone might be broken. A cut might be bleeding badly.

If someone has a broken bone, a doctor has to make sure that it heals right. The bone must be set straight. Then it will heal itself. The doctor puts a hard plaster or plastic **cast** on the arm or leg. The cast keeps the bone straight and protects the injury.

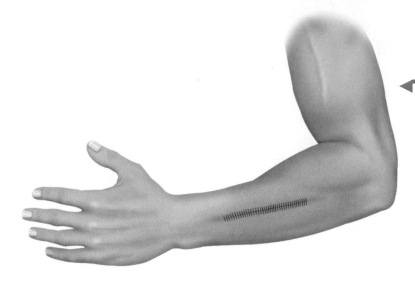

Stitches help skin to heal in the right way.

If a cut is deep, it bleeds a lot. The doctor uses special sewing thread to repair the injury. The doctor puts **stitches** in the skin to pull it back together. This stops the bleeding. When the skin heals, the doctor takes out the stitches. Sometimes it takes a few weeks for a cut to heal.

Most people can go home after they are treated in an emergency room. But some people might have to be moved to a hospital room for more treatment.

CHECKPOINT

1. What rules should you follow during an emergency?
2. What rules should you follow when making an emergency telephone call?
3. Describe what happens in an emergency room.

 What can be done to help someone who is injured?

ACTIVITY

Dialing 911 for an Emergency

Find Out

Do this activity to see what to do during an emergency.

Process Skills

Communicating
Experimenting

WHAT YOU NEED

two old telephones (not hooked up)

Activity Journal

pencils

paper

WHAT TO DO

1. Work with a partner. One partner pretends to dial 911 and pretends to have an emergency.

2. The other partner acts as a 911 operator. She or he writes down the emergency information.

3. The caller gives her or his name, tells what happened, and explains where she or he is.

4. The operator tries to make sense of the information. Can he or she understand the information? What happens when the caller talks too fast? What happens when the caller talks slowly?

5. Partners should change places and practice with a different emergency.

CONCLUSIONS

1. When people are excited and scared, are they calm?

2. Why is it important to stay calm in an emergency?

ASKING NEW QUESTIONS

1. Who answers 911 calls?

2. What does the person who answers a 911 call do if he or she cannot understand the caller?

SCIENTIFIC METHODS SELF CHECK

✔ As the caller, did I **communicate** the information about the emergency to the operator clearly and calmly?

✔ As the 911 operator, did I **collect data** by **writing** down all of the information given to me by the caller?

✔ As the 911 operator, did I **experiment** by trying to get the caller to speak clearly and calmly? Was I able to get all of the important information about the emergency from the caller?

Review

Reviewing Vocabulary and Concepts

Write the letter of the word or phrase that completes each sentence.

1. People who have poor safety habits are more likely to get ___.
   a. emergencies   b. antibiotics
   c. injuries   d. ambulances

2. One piece of safety equipment designed especially for use in a car is the ___.
   a. seat belt   b. helmet
   c. ambulance   d. cast

3. A serious event that comes without warning and calls for fast action is ___.
   a. an ambulance   b. an emergency
   c. an antibiotic   d. a cast

4. ___ helps a broken bone heal in the right way.
   a. A cast   b. A helmet
   c. An ambulance   d. A sprain

Match each definition on the left with the correct term.

5. a piece of safety equipment designed to protect your head   a. first aid

6. care given right away in an emergency   b. helmet

7. an injury to a ligament near a joint   c. stitches

8. one method a doctor or nurse uses to close a cut   d. sprain

Understanding What You Learned

Write the letter of the word or phrase that completes each sentence.

1. Before putting first aid cream on a cut, you should ___.
   a. wash it with soap and water
   b. put a bandage on it
   c. get stitches
   d. call 911

2. To slow the bleeding in a deep cut, you can put pressure on the cut and ___.
   a. wear a knee pad
   b. fasten your seat belt
   c. elevate it
   d. hold your breath

3. If you sprain your ankle, you should ___.
   a. try to run
   b. put first aid cream on it
   c. put pressure on it
   d. rest it and elevate it

4. If someone is hurt badly and first aid is not enough to help, he or she may need ___.
   a. emergency care
   b. knee pads
   c. an air bag
   d. safety goggles

Applying What You Learned

1. Name four types of safety equipment that you should wear to ride a bike or go in-line skating.

 2. Why is it important to know how to practice safe habits and respond to emergencies?

For Your **Portfolio**

Start a list of things people do that might cause an injury. Carry your list with you for one day and add to it whenever you see a safety problem. Next to each item on the list, note a safety practice or a piece of safety equipment that might keep people from getting hurt.

NUTRITION

Have you ever thought about how the food you eat makes you feel? Food affects how you look, feel, and grow. Food gives your body the energy it needs to work and play.

Most people like some foods more than other foods. Did you know that the foods you choose to eat can affect how you feel each day? Some foods help your body stay healthy better than other foods do. Healthful foods help your body repair itself if you are injured or sick. They also supply energy to your body. Eating healthful foods is one way to stay healthy.

The Big IDEA

Nutrients in food support growth and good health.

CHAPTER SCIENCE INVESTIGATION

Use food labels to help you choose which foods to purchase in the grocery store. Find out how in your *Activity Journal.*

Nutrients in the Basic Food Groups

Find Out

- What nutrients are
- What nutrients the bread, cereal, rice, and pasta group gives your body
- What nutrients the fruit group gives your body
- What nutrients the vegetable group gives your body
- What nutrients the meat, poultry, fish, dried beans, eggs, and nuts group gives your body
- What nutrients the milk, yogurt, and cheese group gives your body

Vocabulary

nutrients
vitamin
mineral
protein

The Big QUESTION

Why are some foods better for you than others?

Did you eat healthful foods today? Healthful foods help your body by supplying it with important nutrients. They help your body systems stay strong and healthy.

Nutrients

How do you know which foods are more healthful than others? Foods have nutrients in them. **Nutrients** (no͞o′ trē ənts) are substances your body needs to stay healthy.

Nutrients help the body in three main ways. They provide energy, build and repair body tissues, and control body activities. There are six kinds of nutrients: carbohydrates, proteins, fats, vitamins, minerals, and water.

All foods belong to a food group. There are five food groups. The foods in each food group provide many different nutrients. You should have food from all of the food groups each day.

Bread, Cereal, Rice, and Pasta Group

Foods from the bread, cereal, rice, and pasta group provide energy for your body. Corn, wheat, rice, and oats come from plants called grains. You need 6 to 11 servings of foods from the bread, cereal, rice, and pasta group each day. Healthful foods from this group include pasta, cereal, whole-grain bread, tortillas, and rice. These foods give your body vitamins, carbohydrates, minerals, and some protein.

Some foods from the bread, cereal, rice, and pasta group

Fruit Group

What foods from the fruit group do you like to eat? Fruits contain many vitamins. A **vitamin** is a nutrient found in plant and animal foods that the body needs for growth. Vitamins are also made into tablet form. Some of you may take a vitamin tablet at home. Oranges and strawberries are especially good sources of vitamin C. It is a good idea to eat a variety of fruits because they contain many vitamins. You should eat two to three servings of fruit each day.

Some foods from the fruit group

Vegetable Group

What foods from the vegetable group do you like to eat? Vegetables are a good source of vitamins, minerals, and carbohydrates. Carrots, sweet potatoes, and spinach are good sources of vitamin A. Broccoli and cauliflower are good sources of vitamin C. Broccoli is also a good source of the mineral calcium. A **mineral** (min′ ər əl) is a nutrient your body needs in small amounts to help control body activities. Vegetables are low in fat and also contain water and fiber, which help your body stay healthy. You should eat three to five servings of vegetables each day.

Some foods from the vegetable group.

Meat, Poultry, Fish, Dried Beans, Eggs, and Nuts Group

Meat, poultry, fish, eggs, nuts, and beans are all good sources of **protein** (prō′ tēn). These foods help you stay strong. Your body uses the protein in these foods to grow and to repair itself. Foods in this group also contain B vitamins and minerals. The foods in this group are from plants and animals.

Some foods from the meat, poultry, fish, dried beans, eggs, and nuts group

Chicken, turkey, beef, salmon, eggs, and tuna all come from animals. These foods are all good sources of protein. Soybeans, black beans, almonds, and sesame seeds come from green plants. These foods are also good sources of protein. It is a good idea to eat a variety of meat, poultry, fish, beans, and nuts. You should eat two to three servings of foods from this group each day.

Milk, Yogurt, and Cheese Group

Foods in the milk, yogurt, and cheese group help you have strong bones and teeth because they are good sources of protein and the minerals calcium and phosphorus. Milk and milk products also contain a B vitamin known as riboflavin. Eating three to four servings of foods made with milk each day can help you get enough of these nutrients.

Some foods from the milk, yogurt, and cheese group

Fats, Oils, and Sweets

Some foods such as candy, desserts, dips, and sweet drinks are not healthful. These foods do not have many nutrients. They have extra sugars and fats that your body does not need. Eat these foods only once in a while or in small amounts.

Oils are made of fat and are often used in cooking. While a certain amount of fat in your diet helps your body stay healthy, eating too much fat can cause health problems. Most people get enough fat in their diets just by eating healthful foods. Learn to find the fat content on food labels and avoid eating foods with high fat contents.

	Carbohydrates	Protein	Vitamins	Minerals	Water	Fats
Bread, rice, and pasta group	★	★	★	★		
Fruit group	★		★	★	★	
Vegetable group	★		★	★	★	
Meat, poultry dried beans, eggs, and nuts group		★	★	★		★
Milk, yogurt, and cheese group		★	★	★	★	★

This table shows which food groups are the best sources of the six kinds of nutrients. It is important to eat foods from each of the food groups to get all the nutrients you need.

CHECKPOINT

1. What are nutrients? What kinds of nutrients are there?

2. What kinds of nutrients does cereal give your body?

3. What kinds of nutrients do oranges give your body?

4. What kinds of nutrients does broccoli give your body?

5. What kinds of nutrients does meat give your body?

6. What kinds of nutrients does yogurt give your body?

? Why are some foods better for you than others?

ACTIVITY
Reading Food Labels

Find Out

Do this activity to learn how to find the fat content of foods by reading nutrition labels.

Process Skills

Observing
Communicating

WHAT YOU NEED

one or two bags of food (chips, cookies, crackers, candy, or other snacks)

box of food (cereal, pasta, or rice)

can of food (beans, soup, fruit, or vegetables)

Activity Journal

WHAT TO DO

1. **Observe** each of the foods, reading the labels on their packaging. Read the "Nutrition Facts" label and find the total fat grams in one serving of each food.

2. **Record** the name of each food and the total fat grams in each serving of that food. (A serving is the amount of a food that is normally eaten at one time.)

3. Repeat Step 2 for each of the foods you observed during the activity. Compare the fat contents of the foods to one another. Which of the foods had the highest fat contents?

CONCLUSIONS

1. Which types of food contained large amounts of fat?

2. What else did you see on the Nutrition Facts labels?

ASKING NEW QUESTIONS

1. How else could you find out which foods contain large amounts of fat?

2. Why is it important to learn to read nutrition labels?

SCIENTIFIC METHODS SELF CHECK

✔ Did I **observe** the Nutrition Facts labels and read the fat content for a single serving of each food?

✔ Did I **record** my observations?

The Benefits of Good Nutrition

Find Out

- What a balanced diet is
- How carbohydrates help your body
- How proteins help your body
- How vitamins help your body
- How minerals help your body
- Why water is an important nutrient
- Why fat is an important part of a healthful diet

Vocabulary

diet
carbohydrates

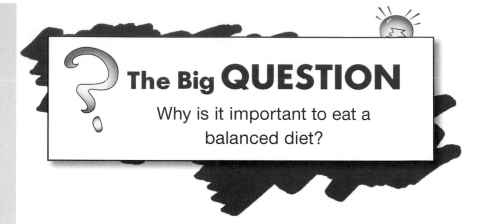

? The Big QUESTION

Why is it important to eat a balanced diet?

Your body needs nutrients to work well each day. How does your body get the nutrients it needs? It breaks down the food you eat into carbohydrates, proteins, fats, vitamins, minerals, and water. A balanced diet gives your body the nutrients it needs to grow and stay healthy.

Eating a Balanced Diet

How do you plan a balanced diet? First, it is important to understand what a diet is. A **diet** is all the foods you usually eat and drink each day. Some people may have to follow a special diet for medical reasons. Make a list of all the foods you eat during one week. The list is your diet for that week.

A balanced diet includes foods from each of the five food groups. Eat a variety of foods from the food groups to get all the nutrients you need. Eating a balanced diet helps your body in many ways.

Preparing a balanced meal

Foods with lots of carbohydrates

Carbohydrates

Carbohydrates (kär′ bō hī′ drāts) are your body's most important source of energy. Having energy is having the strength to do active things without getting tired quickly. Carbohydrates are found most often in foods from plants. Carbohydrates include sugars, starches, and fibers. Sugars and starches give your body energy. Fiber in the diet helps keep the digestive system healthy. Carbohydrates also help your nerves and blood cells. Good sources of carbohydrates are vegetables such as peas and potatoes, pasta, seeds, tortillas, brown rice, whole-grain breads, and cereals. Which carbohydrates are in your diet?

Proteins

Proteins are important to many parts of your body. Your muscles, bones, teeth, skin, blood, and organs such as your heart all contain protein. Your body uses protein to grow and repair body tissues. As you grow, proteins help your body build new body tissues. If you get hurt, protein helps your body build new tissues to repair the injury. Protein comes from both plant and animal foods. Fish, meat, chicken, turkey, and eggs are all good sources of animal protein. Beans, nuts, tofu, and seeds are all good sources of plant protein. Dairy products also contain protein. As you grow and play, protein from your diet helps you get stronger.

The protein in your food helps you get stronger.

These foods contain many of the vitamins your body needs to work well.

Vitamins

Vitamins help your body processes work well. You get vitamins in your diet by eating a variety of foods. Your body needs different vitamins for different reasons. Vitamin D works with calcium to develop strong bones. Vitamin A helps your bones grow and helps keep your eyes healthy. Vitamins E and K help protect your blood cells. Vitamin C helps your body fight infections. The B vitamins help many parts of your body, such as your brain, muscles, nerves, and blood. Cooking fruits and vegetables too long can remove many of the vitamins. Eat fruits and vegetables raw or slightly cooked to get the right amount of vitamins from them.

Minerals

You get minerals from the foods you eat and drink. You get calcium and phosphorus from leafy greens, broccoli, beans, and dairy products. Your body uses the calcium and phosphorus in foods to build your bones and teeth. As you grow, eating foods with these minerals helps keep your bones and teeth hard. Building up stores of calcium in your bones while you are young will help strengthen them for a lifetime.

Water

Did you know that you get some minerals from drinking water? Drinking water contains the minerals sulfur, chlorine, sodium, calcium, magnesium, potassium, and sometimes fluorine.

Water is the body's most important nutrient. Without water, you could not live. Water is important for every body function. It carries nutrients throughout your body. It is the main part of blood. About two-thirds of your body is made up of water. Try to drink about six to eight glasses of water each day. Foods also supply some water. Fruits, vegetables, and milk are some of the foods that contain a lot of water.

Healthy teeth depend on foods with minerals.

Fats

Fats are another source of energy. Almost every food contains some fat. Proteins such as meat, chicken, eggs, nuts, and fish all have some fat. Milk products contain fat. You also get fat from cheese, butter, and oils. Fats take longer to break down than other foods, so they keep you from getting hungry. Fats help cushion the bones in your body's joints. They are also needed by your body to keep your nerves and blood healthy. Fats carry many important vitamins, such as vitamins A, D, E, and K. Your body needs only a certain amount of fat to stay healthy.

Eating a balanced diet can help you grow and stay healthy. Try to eat three meals a day and at least two healthful snacks. If you choose foods from all the food groups, you will have a balanced, healthful diet.

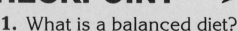

Many foods with fats contain vitamins that help keep your skin healthy.

CHECKPOINT

1. What is a balanced diet?
2. Why does your body need carbohydrates?
3. Why does your body need proteins?
4. Why does your body need vitamins?
5. Why does your body need minerals?
6. Why is water an important nutrient?
7. Why is fat an important part of a healthful diet?
   ❓ Why is it important to eat a balanced diet?

ACTIVITY

Finding Water in Fruits

Find Out

Do this activity to learn how to tell if a food contains water.

Process Skills

Observing
Hypothesizing
Communicating
Predicting

WHAT YOU NEED

fruit for slicing (apple, pear, plum, peach)

paper clips

hot plate (optional)

knife

simple balance scale

Activity Journal

WHAT TO DO

Safety! *The teacher should cut the fruit into thin slices as you observe.*

1. Your teacher will cut the fruit slices. **Observe** the fruit. Do the slices look wet? Do they feel wet?

2. **Make a hypothesis.** What makes the fruit slices look and feel wet?

3. Put a fruit slice on one side of the balance. Put paper clips on the other side. **Record** the number of paper clips you need to balance the slice.

4. Put the slice in a sunny place. Leave it out overnight. Or, your teacher may put the apple slice on a hot plate for about an hour.

5. **Predict** how the fruit slice will change. **Record** your prediction.

6. **Observe** the fruit slice. **Record** your observations.

7. Put the fruit slice on the balance again. Use paper clips to balance the fruit slice. How many paper clips do you need now?

8. Repeat this activity using the same kind of fruit. Compare your first and second measurements.

9. Repeat Steps 2 through 8 using a different kind of fruit. Again compare your first and second measurements.

CONCLUSIONS

1. Compare your prediction with your observation.

2. How did the fruit slices change?

3. What was removed from the fruit slices? How do you think it was removed?

4. Were your first and second measurements the same or different?

ASKING NEW QUESTIONS

1. Raisins are dried grapes. How could you make raisins?

SCIENTIFIC METHODS SELF CHECK

✔ Did I **make a hypothesis** about the appearance of the fruit slices?

✔ Did I **record** the number of paper clips needed to balance the fruit slices?

✔ Did I **predict** how the fruit slices would change?

✔ Did I **observe** the fruit slices and **record** my observations?

Review

Reviewing Vocabulary and Concepts

Write the letter of the word or phrase that completes each sentence.

1. Carbohydrates, proteins, fats, vitamins, minerals, and water are six kinds of ___.
   - **a.** energy
   - **b.** nutrients
   - **c.** diet
   - **d.** fruit

2. Your body uses ___ to grow and repair itself.
   - **a.** proteins
   - **b.** sweets
   - **c.** a food group
   - **d.** water

3. All of the foods you usually eat and drink each day make up your ___.
   - **a.** fiber
   - **b.** diet
   - **c.** carbohydrates
   - **d.** fats

4. ___ are the body's most important source of energy.
   - **a.** Fruits
   - **b.** Minerals
   - **c.** Vitamins
   - **d.** Carbohydrates

Match each definition on the left with the correct term.

5. the most important nutrient
   - **a.** calcium

6. mineral that helps keep teeth and bones strong
   - **b.** protein

7. a nutrient found in meat, poultry fish, nuts, beans, tofu, and seeds that helps your body build new tissues if you get hurt
   - **c.** nutrients

8. substances your body needs to stay healthy
   - **d.** water

Understanding What You Learned

1. List the five food groups.

2. What nutrients are in bread?

3. What nutrients are in fruits and vegetables?

4. What nutrients do you find in meat?

Applying What You Learned

1. Why are some foods better for you than others?

2. What is a balanced diet, and what is one way to be sure that you have a balanced diet?

3. Explain what minerals and fats do for your body.

 4. Name two good things that nutrients do for your body.

For Your Portfolio

Think about the five food groups and the variety of foods that belong to those groups. Plan a menu for a balanced meal. Write down your menu. Draw a place setting on your table. Label the foods. Then list the foods and the nutrients found in each.

Unit Review

Concept Review

1. Describe how your brain, muscles, and bones work together in your body.

2. What safety habits are important for sports and other activities, and what can you do if you have an accident?

3. What foods are important for you to eat and what do they do for your health?

Problem Solving

1. Describe the muscles and bones in your favorite animal or pet and how they help it move.

2. If you were ice skating with a friend, what could you do to stay safe? What would you do if your friend got hurt?

3. Describe a healthful menu for lunch.

Something to Do

With a group, write a travel plan for a camping or hiking trip. On one page, list the muscle groups that you think would get stronger from your trip. On a second page, list the first aid supplies you would bring with you, and include what you might do in an emergency. On the third page, list the foods you would bring with you on your trip.

Reference

Animal Adaptations

Ptarmigan in summer and in winter
Ptarmigans' feathers change with the seasons, providing the birds with excellent camouflage in the snow during winter and in the grasses, brush, and soil during the warmer seasons.

Puffer fish inflated and normal size
When threatened, the puffer fish gulps large amounts of water, causing its body to become much bigger. This larger size makes it seem more threatening to its predators, frightening many of them away.

Arctic hare and jackrabbit
The Arctic hare has adapted to a cold habitat, while the jackrabbit has adapted to a hot habitat. The Arctic hare's shorter ears, more compact body, and thick coat all help it to keep heat in its body. The jackrabbit's longer ears, slimmer body, and thin coat help it to stay cool in the heat of warm desert habitats.

Emperor penguin
Emperor penguins have adapted to life in cold habitats. They have a thick layer of fat underneath their skin to help them stay warm in icy environments. They also have streamlined bodies, allowing them to move quickly and gracefully in the water as they search for food.

Sidewinder

The sidewinder has adapted to its sandy desert habitat. To move quickly across the hot, shifting sand, sidewinders throw their bodies into the air in repeated sideways leaps instead of moving in one smooth motion.

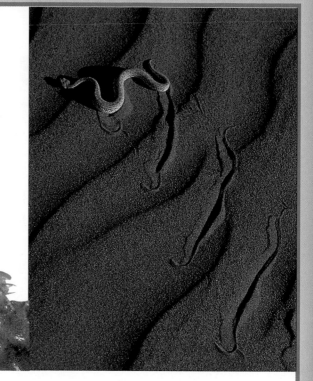

Poison-dart frog

The poison-dart frog is one of the most poisonous animals in the world. Their bright colors warn predators that their skin contains deadly poisons, keeping many predators away.

Fennec fox

The desert-dwelling fennec fox has huge ears that allow heat to escape its small body easily during high daytime temperatures. The fennec fox sleeps in a cool underground den or cave during the day, where it is protected from the hot desert sun. As the sun sets, desert temperatures drop. The fennec fox leaves its burrow to search for food, protected from the wind and cold of the night by its furry coat.

Porcupine

The porcupine has a defense adaptation that protects it from predators. A porcupine can point the long, sharp quills on its body toward a predator by turning its back, shaking its body, and rattling its quills in warning. If a predator tries to touch the porcupine, the quills easily come loose, sticking into the predator's body, and frightening it away.

Leaf-tailed gecko

The leaf-tailed gecko uses camouflage to hide from predators in its forest home. When a gecko is holding still, it is camouflaged against the trunks of certain trees, making it nearly impossible to see.

Endangered Animals

A threatened species is one that could become endangered in the near future. An endangered species is one that could become extinct in the near future. Some endangered animals are extinct in the wild and now live only in zoos.

Giant Panda
(China)

Threatened Animals

This chart shows the number of threatened species in each animal group.

Animal Group	Number of threatened species
Mammals	612
Birds	704
Reptiles	153
Amphibians	75
Fish	434
Insects	377

Queen Alexandra's Birdwing
Butterfly (New Guinea)

Red Ruffed Lemur
(Madagascar)

Biggest Threats to Wildlife

Humans use up land and resources, destroying animals' natural habitats.
Animals are killed for their skins, bones, and horns.
Pollution poisons animals.

Some of the Most Endangered Animals

Whooping Crane
(Texas coast)

Golden Lion Tamarin
(Brazil)

Snow Leopard
(Central Asia)

Asian Elephant
(Southeast Asia)

Woolly Spider Monkey
(Brazil)

Mountain Gorilla
(Central Africa)

Kemp's Ridley Sea Turtle
(Mexico)

Solar System

Mars Facts

Diameter	6,787 km
Surface temperature	-143 to 17°C
Distance from sun	228 million km
Time taken to circle sun (year)	687 Earth days
Time taken to turn on axis (day)	24 hours, 37 minutes

Pluto Facts

Diameter	2,300 km
Surface temperature	-233 to -223° C
Distance from sun	5,900 million km
Time taken to circle sun (year)	248 Earth years
Time taken to turn on axis (day)	6 Earth days, 9 hours

Saturn Facts

Diameter	120,660 km
Surface temperature	-178° C
Distance from sun	1,427 million km
Time taken to circle sun (year)	29.46 Earth years
Time taken to turn on axis (day)	10 hours, 12 minutes

Mercury Facts

Diameter	4,878 km
Surface temperature	-173 to 427° C
Distance from sun	58 million km
Time taken to circle sun (year)	88 Earth days
Time taken to turn on axis (day)	58.6 Earth days

Earth Facts

Diameter	12,756 km
Surface temperature	−89.6 to 58° C
Distance from sun	150 million km
Time taken to circle sun (year)	365.26 days
Time taken to turn on axis (day)	23 hours, 56 minutes

Neptune Facts

Diameter	49,528 km
Surface temperature	-214° C
Distance from sun	4,497 million km
Time taken to circle sun (year)	164.8 Earth years
Time taken to turn on axis (day)	19 hours, 6 minutes

Uranus Facts

Diameter	51,118 km
Surface temperature	-216° C
Distance from sun	2,870 million km
Time taken to circle sun (year)	84 Earth years
Time taken to turn on axis (day)	17 hours, 54 minutes

Jupiter Facts

Diameter	142,800 km
Surface temperature	-148° C
Distance from sun	778.3 million km
Time taken to circle sun (year)	11.86 Earth years
Time taken to turn on axis (day)	9 hours, 55 minutes

Venus Facts

Diameter	12,104 km
Surface temperature	462° C
Distance from sun	108 million km
Time taken to circle sun (year)	225 Earth days
Time taken to turn on axis (day)	243 Earth days

Global Landforms

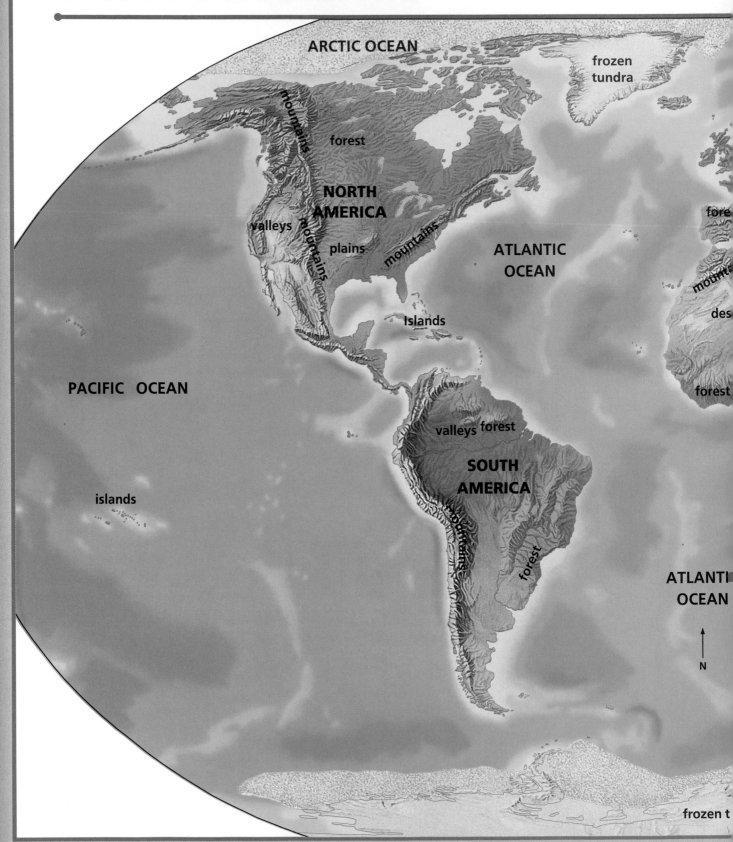

ARCTIC OCEAN

frozen tundra

mountains

forest

NORTH
AMERICA

valleys

mountains

plains

mountains

ATLANTIC
OCEAN

Islands

PACIFIC OCEAN

fore

mounta

des

forest

islands

valleys forest

SOUTH
AMERICA

mountains

forest

ATLANTI
OCEAN

N

frozen t

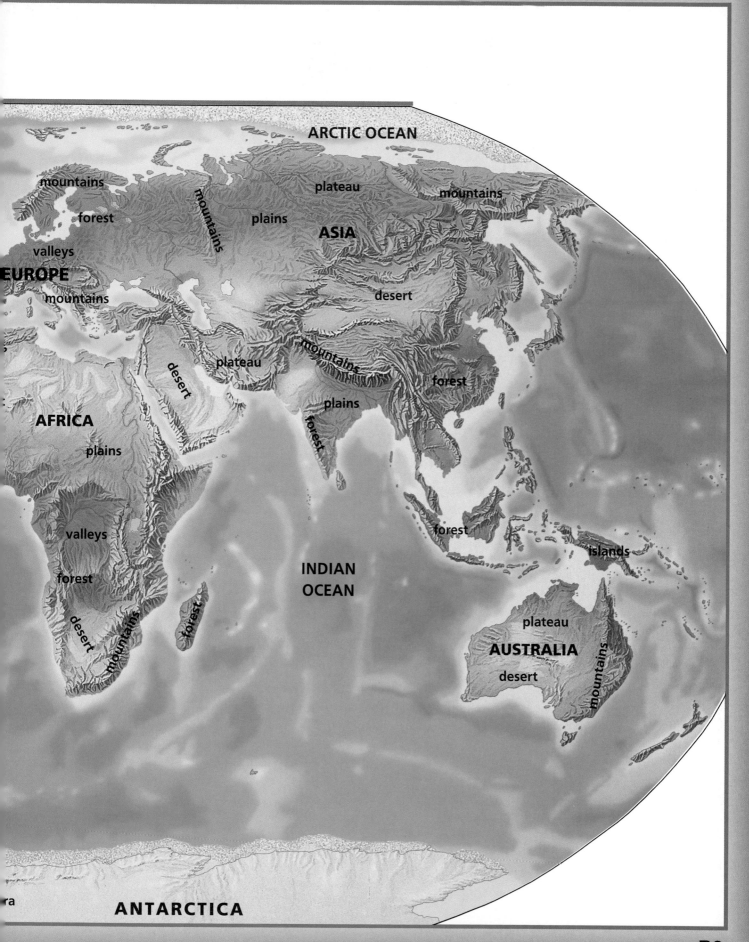

Electromagnetic Spectrum

Electromagnetic energy travels through space in waves. All these energy waves together make up the electromagnetic spectrum. Light is one form of electromagnetic energy. It is the only form we can see.

long wavelength (more than 1 km)

(waves not to scale)

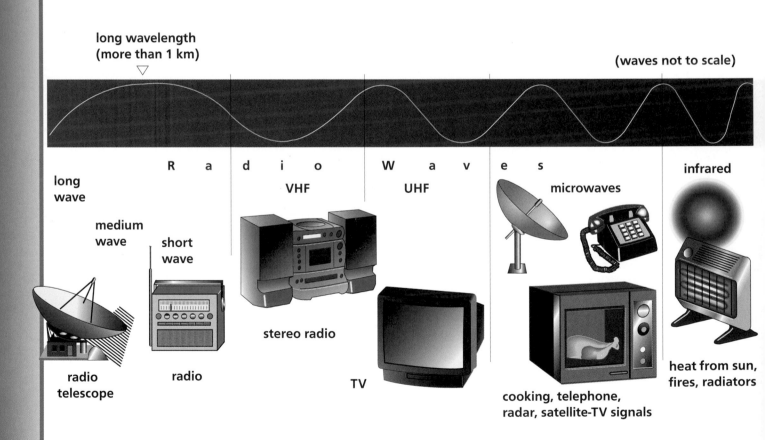

R a d i o   W a v e s

long wave

medium wave

short wave

VHF

UHF

microwaves

infrared

radio telescope

radio

stereo radio

TV

cooking, telephone, radar, satellite-TV signals

heat from sun, fires, radiators

Some electromagnetic waves are very long—
thousands of meters long. Some waves are
very short—just a fraction of a millimeter
long. The shorter the wavelength, the more
energy the wave carries.

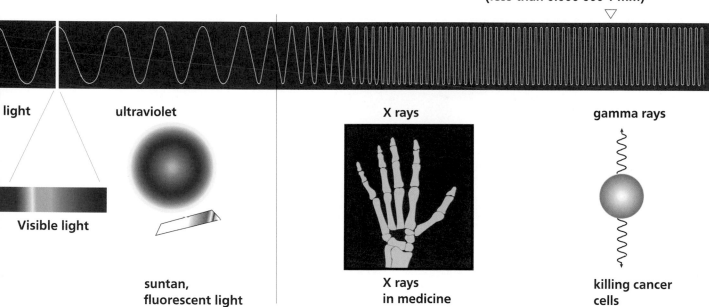

**short wavelength
(less than 0.000 000 1 mm)**

light

Visible light

ultraviolet

suntan,
fluorescent light

X rays

X rays
in medicine

gamma rays

killing cancer
cells

Heat Transfer

The sun is the most important source of energy on Earth. The sun's energy travels 150 million kilometers to Earth. Energy from the sun is converted to heat and light. Heat is a very important form of energy called *thermal energy*. When thermal energy moves from one area or object to another, the energy is *transferred*. Three ways that thermal energy is transferred are conduction, convection, and radiation.

Conduction is the transfer of thermal energy through solid objects. When one solid object (such as a metal pan) touches a hotter solid object (such as an electric burner), heat moves from the warmer to the cooler object. Materials that conduct heat well, (such as metal), allow thermal energy to pass through evenly, making them good materials to use for cooking. When a metal pot heats up, its contents heat up. The heat source and all conducting material touching it will transfer thermal energy.

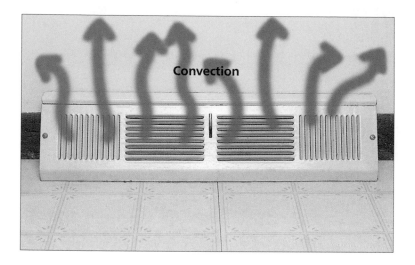

Convection is the transfer of thermal energy through the movement of heated liquids and gases. Heat transfer by convection always involves the movement of matter, which carries the heat along as it moves.

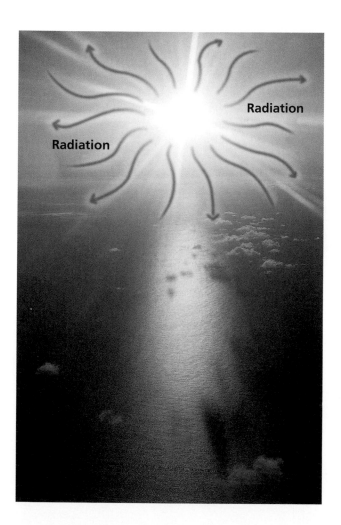

Radiation is a way of transferring thermal energy that does not involve any matter at all. Both conduction and convection use matter to move thermal energy from one place to another, but radiation is different. Radiation uses electromagnetic waves, not matter, to move heat. The heat energy that comes to Earth from the sun travels through space by radiation.

Bone Systems

Skeletal System

The skeletal system is all your bones. There are 206 bones in the human body. Your bones give you support and protect your organs.

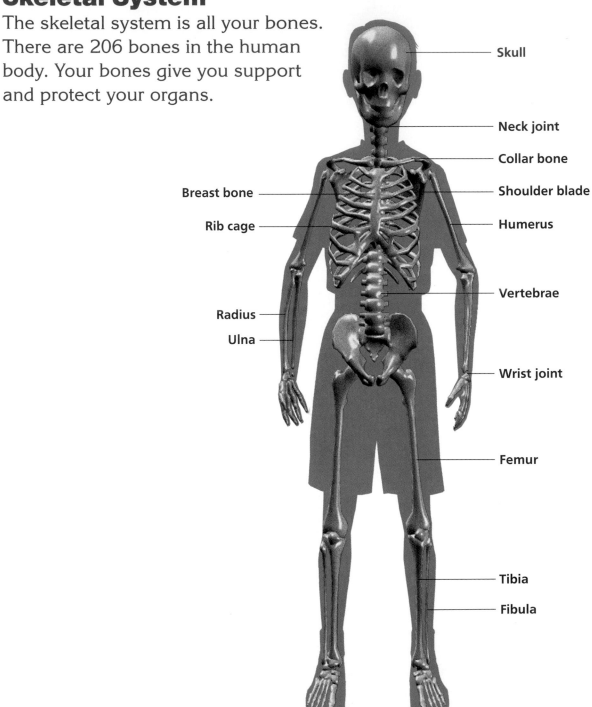

- Skull
- Neck joint
- Collar bone
- Breast bone
- Shoulder blade
- Rib cage
- Humerus
- Vertebrae
- Radius
- Ulna
- Wrist joint
- Femur
- Tibia
- Fibula

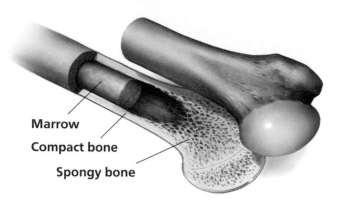

Marrow
Compact bone
Spongy bone

Inside a Bone

Fast Facts

- Your longest bone is your thigh bone. It's called the femur.
- Babies have more than 300 bones. Some of the bones grow together later.
- Humans and giraffes have the same number of neck bones.

Types of Joints

The point at which two bones meet is called a joint. Different joints in the body help bones move in different ways.

The shoulder is a ball-and-socket joint. Such a joint allows for a complex and large range of motion.		
The elbow is a hinge joint. A hinge joint allows movement in one direction like a door hinge.		
Your topmost vertebra forms a pivot joint with the vertebra beneath it that allows you to move your head from side to side, as though pivoting around an axis.		
Joints in your wrist are gliding joints. Movement in gliding joints is quite complex.		

Muscular System

The muscular system is all your muscles. You have more than 600! Most muscles pull your bones so you can move. Some muscles work inside your organs. Your heart is a muscle.

Muscles pull, but they can't push. They have to work together to move your bones. Take your arm, for example: One muscle (a biceps) pulls your forearm up to bend your arm. To straighten your arm, another muscle (a triceps) has to pull the forearm down.

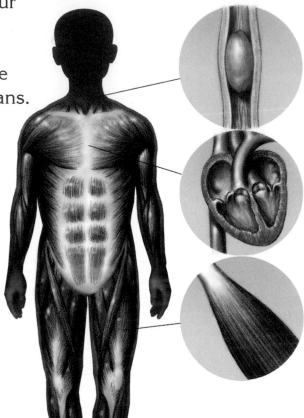

smooth muscle

cardiac muscle

skeletal muscle

Fast Facts

- The smallest muscle is in your ear. It is only about 1 mm long. Your biggest muscle is called the gluteus maximus. It's a muscle in your buttocks and thighs.
- Your body weight is about 40% muscle.
- It takes 43 muscles to frown, but only 17 to smile.

Home Safety

People have more accidents at home than they do anywhere else. Here are some home safety tips.

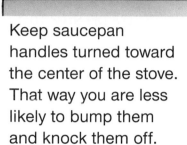

Do not use any electrical equipment in or near water.

Keep saucepan handles turned toward the center of the stove. That way you are less likely to bump them and knock them off.

Store cleaning products and other dangerous items on a high shelf where children can't reach them. Do not put dangerous substances in familiar food or drink containers. Put big labels on all dangerous substances.

Run cold water in the tub first. Then keep testing it as you add hot water. Do not put your hand under the hot water when it's running.

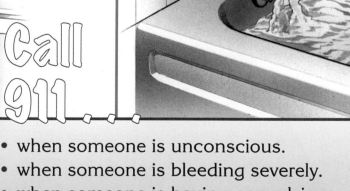

Call 911...

- when someone is unconscious.
- when someone is bleeding severely.
- when someone is having convulsions.
- when someone is having trouble breathing.
- when someone has severe stomach pains.

The Food Guide Pyramid

A Guide to Daily Food Choices

Giving your body all of the nutrients it needs helps it to defend itself against disease. Good nutrition also helps strengthen and maintain developing body parts. The Food Guide Pyramid shows the number of recommended daily servings of each of the basic food groups. Eating a variety of foods from each of the basic food groups each day will help you get all of the nutrients you need to stay healthy. Note that there are no recommended servings of fats, oils, and sweets, as these foods should be eaten only occasionally and in small amounts.

Fats, Oils, and Sweets
Use Sparingly

Milk, Yogurt, and Cheese Group
2–3 servings

Meat, Poultry, Fish, Dry Beans, Eggs, and Nuts Group
2–3 servings

Vegetable Group
3–5 servings

Fruit Group
2–4 servings

Bread, Cereal, Rice, and Pasta Group
6–11 servings

OIL

MILK

Yogurt

Flour

Vitamins and Minerals

Vitamin	Why We Need It	Where to Get It
A	Healthy eyes and skin; growth; helps fight infection	liver, egg yolks, yellow and orange fruits and vegetables
B1	Healthy nervous and digestive systems	whole-grain breads and pastas, brown rice, liver, beans, peas, nuts
B2	Keeps tissues healthy	milk, liver, cheese, eggs, green vegetables, lean meat
B3	Energy; healthy skin	liver, pork, poultry, fish, whole-grain breads and cereals, nuts, dried beans
B6	Producing red blood cells	liver, pork, poultry, fish, bananas, potatoes, most fruits and vegetables
C	Healthy skin, teeth, bones, and tissues; fighting disease	citrus fruits, strawberries, tomatoes, potatoes
D	Strong teeth and bones	salmon, liver, eggs, sunlight
E	Producing red blood cells; protecting the lungs	margarine, lettuce, leafy green vegetables, whole-grain cereals, nuts

Mineral	Why We Need It	Where to Get It
Calcium	Healthy bones and teeth; healthy nerve cells; helps blood clot	milk, cheese, yogurt, fish, bread, green vegetables
Iron	Helps build healthy blood, strengthen muscles, fight infections	red meat, beans, egg yolks, dried fruit, nuts, grains, spinach
Potassium	Keeps tissues healthy; aids nerve impulse function	oranges, bananas, dry beans, molasses
Magnesium	Helps build strong bones and teeth	vegetables, meat, leafy greens, potatoes

Glossary

A

adapt to change size, shape, or features over time to adjust to a changing environment

adaptations changes in structures or behaviors that help living things stay alive

atom a very small piece of any kind of matter or element

B

behavior how a living thing acts in its environment

biodegradable things that will decay or break down naturally

C

calcium a mineral used by the body and stored in bones

camouflage the coloring, shape, and size that help some animals blend in with or look like their environment

carbohydrates nutrients found in food from plants that give your body energy

carbon dioxide a colorless, odorless gas made of carbon and oxygen; plants use it to grow

cardiac muscle the type of muscle that makes up the heart

cast a hard plaster or plastic shell put on an injured limb to keep the bone straight

charge an amount of electricity—all matter and living things have tiny charges

circuit the pathway that electrical energy moves through

clay soil soil that has tiny grains in it, feels smooth, holds water well, and can get hard and packed

community all the living things (animals, plants, and other organisms) that live and interact in an area

compost a mixture of biodegradable materials that can enrich soil

Glossary

compound two or more atoms of different elements joined to make matter

conductors materials that conduct electric current easily

conservation when people work to protect Earth and its resources

constellations groups of stars that form a pattern

consumers organisms that do not make their own food; they eat other plants and animals

contracting when a muscle pulls together and gets shorter

core the center of Earth, which is made of two layers—the inner core and outer core

cornea a clear outer layer at the front of the eyeball that lets light enter our eyes

crescent moon the phase in which we can see a thin sliver of the lighted side

crust the upper layer of Earth

D

decomposers organisms that get food by breaking down dead plant and animal matter

deposition when running water, ice, or wind moves and deposits small rock particles, creating landforms

diet all the foods you usually eat and drink each day

distance how far something moves

E

earthquake the sudden movement of Earth's crust

eclipse when an object passes into the shadow of another object

ecosystem groups of living and nonliving things and their habitats

effort the force applied to provide the energy to move a load

Glossary

electric charges forces found in all matter and living things

electric current the electrical energy that flows through a circuit

electrical energy energy that flows through wires to make things work

electromagnet a magnet with a core of iron or steel inside wire coils that carry electric current

element matter that is made up of only one kind of atom

emergency care given when an injured person needs special help right away

endangered living things that are in danger of becoming extinct

energy the ability to make things move or change

environment all of the living and nonliving things that surround an organism

environmentalists people who work to protect the environment by using conservation

evaporation the change from a liquid to a gas

extinct when every kind of a certain plant or animal has died

F

faults deep cracks in Earth's crust along which movement can occur

first aid the care given right away for an injury

first-quarter moon when the moon has moved one quarter through its orbit and we see half of the side that faces us

food chain plants, animals, and other organisms feeding upon each other and passing energy on from one organism to another

Glossary

food web the different overlapping food chains in a community linked together

force the effort made when an object is pushed or pulled

fossils remains of organisms that lived a long time ago and are now preserved

fuels substances that burn and release heat

fulcrum a pivot point on which a lever rests

G

galaxy a large group of stars and planets; Earth is in the Milky Way galaxy

glacier a mass of ice that flows slowly over land

gravity the force that pulls things toward the surface of Earth

H

habitat the place in an ecosystem where living things live and grow

heat energy moving from a warmer object to a cooler object

helmet special equipment with a hard plastic shell and straps that protects your head from a fall or blow

humus a part of soil that forms from decaying plant and animal material

I

igneous rocks that form when melted minerals cool

inclined plane a flat surface that is higher at one end

inexhaustible cannot be used up

inherited characteristics adaptations that living things inherit from their parents

Glossary

injuries cuts, scrapes, or bruises that can happen while playing or working

involuntary muscles muscles that work without you thinking about them

L

landform a feature of Earth's surface, such as a hill or an island

last-quarter moon when the moon is three-quarters of the way through its orbit and we can see half of the side facing us

learned characteristics characteristics that are taught

lens a part of the eye that is behind the pupil and that focuses the image on the retina

lever a board or piece of strong material that rests or tilts on a fulcrum, or pivot point

ligaments strong tissues that connect bones in a skeleton

load the item to be moved by a lever

loam a mixture of clay, sand, and humus that feels soft and holds water well

M

magnetism the ability of a magnet to exert a force

mantle the layer of Earth between the crust and the core

metamorphic rocks that form when heat and pressure change existing rocks into different kinds of rocks

mineral a nutrient that works like vitamins to help control body processes

minerals nonliving materials from Earth called elements; all rocks are made of one or more minerals

mirror a shiny, flat surface that forms a clear image by reflecting light rays

Glossary

mixture a combination of two or more different substances, where each substance keeps its own properties

mountain any landform that is at least 610 m above the ground around it

muscular system all of the muscles in your body; you have more than 600 muscles

N

natural resources materials found in nature that are useful and necessary to living things

new moon when the moon is between Earth and the sun; the lighted half faces away from Earth

nonconductors materials that do not conduct electric current very well

nonrenewable cannot be replaced in a span of time that is useful for people

nosebleed an injury to the nose that causes it to bleed

nutrients substances that provide energy, build and repair tissues, and control processes in your body

O

opaque an object that does not let light pass through it

orbit the path of each planet as it travels around the sun

organisms living things

oxygen a gaseous element found in the air

P

phases the moon's changing appearance; a result of how much of the lighted side is seen from Earth

plain a wide, mostly flat expanse of land

planet a heavenly body that moves around a star

planetarium a museum where you can learn about stars and planets

Glossary

plateau an area of flat land that is raised above the area around it

population a group of the same kind of living thing in the same area

predator an animal that kills other animals for food

prey the animals that are captured and eaten by a predator

prism a triangular piece of glass or plastic that can bend light

producers organisms that use the sun's energy to make food energy

property something about an object or substance that can be measured or sensed

protein a substance found in all living plant and animal cells; protein keeps your body strong

pupil the round, black spot in the eye that lets light enter the eye

R

recycle to change something that has been used into something that can be used again

refract when a light ray bends as it passes through a prism or clear substance

renewable resource that can be replaced in about 30 years

reproduce when organisms have young of their own kind

resource a material found in nature that is useful or necessary to living things

restore when nature repairs itself after a disaster

retina a place in the lining at the back of the eyeball where images are focused and then sent to the brain

rock cycle the constant changing of rocks because of weathering, erosion, melting, cooling, and heating

Glossary

S

sandy soil soil that has large grains, feels gritty, does not hold water well, and is easy to dig

scatter the way light rays bounce or spread off of a surface or particles in many different directions

scavengers animals that get food by eating dead organisms

seat belts special belts in cars that hold passengers in place

sedimentary a kind of rock made from weathered and eroded pieces of rock that are moved, deposited, and squeezed together which then harden

shadow a dark area caused by the blockage of light

skeletal muscles muscles that move the bones in your skeleton

skeletal system all of the bones in the body

skull the bones in your head

smooth muscle involuntary muscle that moves food through your body

soil tiny bits of broken up rock and rotting plant and animal material; the top layer of Earth's surface

solar energy the energy we get from the sun

solar system everything that travels around the sun

solution a mixture formed when one substance dissolves in another substance

spectrum shows all of the colors that make up visible light

speed a way to measure how fast an object moves over a distance

sprain an injury near a joint; sprains often happen to an ankle, wrist, finger, or knee joint

stars huge balls of hot gases

static electricity a form of electricity in which the electric charges build up and then do not move

Glossary

stitches something doctors use to pull skin together when a cut is deep and bleeds a lot

substances basic materials that make up all things

T

telescope an instrument used to observe objects that are very far away

tendons strong cords of tissue that help muscles move bones

topographic maps maps that show elevation

translucent matter that lets some light pass through it

transparent objects that allow light to pass through them

V

vitamin a nutrient found in foods that the body needs for growth

volcano a landform created by the eruption of liquid rock from under Earth's crust

voluntary muscles the muscles that you control

W

work when a continuously applied force makes an object move over a distance

Index

A

activities

adaptations

Index

Index

Index

heat, C22–C25, C34–C35
in ecosystems, A14–A17
in food chains, A14–A16
in food webs, A14, A16
solar, C32–C35
stored, C20–C21
See also electrical energy; force;
    light; machines

environment
changes in the, A50–A57
definition, A4
protecting the, A40–A45,
    A71–A73

environmentalists, A40–A45

evaporation, C9

extinction
definition, A62
natural, A68–A69
of dinosaurs, A60–A63
of the dodo bird, A64
of the great auk, A64
of the passenger pigeon, A65

eye, human, C38–C39

F

faults, B44
Finding Water in Fruits,
    D74–D75

first aid, D40–D47

activities, D46–D47
cuts, D42–D43
emergency, D48–D55
nosebleeds, D45
sprains, D44

food chains, A14–A16
activities, A18–A19
effects of extinction on, A70

food webs, A14

force
activities, C68–C69
and motion, C64–C65
and work, C65
definition, C62–C63
measuring, C66–C67
See also gravity; machines

fossils, B33
dinosaur, A60–A61

fuel energy, C22–C24

fulcrum, C72–C73

G

galaxy, B20–B21

glaciers, B42
activities, B46–B47

gravity, C63–C64

greenhouse effect, A73
activities, A74–A75

Index

Index

Index

motion, C62–C67
See also force; speed

Mount Saint Helens, A52–A56

mountain(s)
formation of, B42–B43,
B49–B53

Moving Ice, B46–B47

muscle(s)
activities, D18–D19, D26–D27
and bones, D2–D4, D9,
D12–D17
and tendons, D9
cardiac, D24
contraction, D12
growth of, D5
involuntary, D16–D17
kinds of, D24–D25
skeletal, D24
smooth, D25
voluntary, D14–D15

muscular system, D5

N

natural resources
air, A36
conservation of, A40–A45,
B80–B81
definition, A34, B77

inexhaustible, B79
nonrenewable, B78
renewable, B77
soil, A35, B72–B73
solar energy, B79
water, A34
wind, B79

nutrition
activities, D66–D67
carbohydrates, D69
diet, D68–D73
fats, D73
food groups, D61–D65
minerals, D60, D61, D63, D64,
D65, D72
nutrients, D60–D65
proteins, D60, D61, D63–D64,
D65, D70
vitamins, D60, D61, D62, D65,
D71
water, D72

O

Observing a Chemical Change,
C18–C19

**Observing Part of an
Ecosystem,** A10–A11

Observing Soil Types,
B74–B75

Index

Index

Index

Credits

Photo Credits

Cover, Title Page, Unit Openers, ©Tim Davis/Tony Stone Images; Back Cover, ©Renee Lynn/Tony Stone Images; iv (tl), ©Karl & Jill Wallin/FPG International, (bl), ©Corbis-Bettmann; v, ©David Weintraub/Photo Researchers, Inc.; vi (tl), ©NASA/Phototake, (bl), ©David L. Brown/Tom Stack & Assoc.; vii, ©Greg Vaughn/Tom Stack & Associates; viii (cl), ©Stephen Johnson/Tony Stone Images, (bl), ©Barbara Filet/Tony Stone Images; ix (cl), ©Aaron Haupt/Photo Researchers, Inc., (bl), ©Patrick Bennet/Corbis; x (tl), ©Richard Hamilton Smith/Corbis, (bl), ©David Young-Wolff/PhotoEdit; xi, ©David Young-Wolff/PhotoEdit; xii, xiii, xiv, xv, ©Matt Meadows; A2-A3, ©Karl & Jill Wallin/FPG International; A5 (bl), ©Joanna McCarthy/The Image Bank, (br), ©George H. H. Huey/Animals Animals, (tr), ©Peter B. Kaplan/Photo Researchers, Inc.; A6, ©Dan Guravich/Photo Researchers, Inc.; A8, ©Michael Newman/PhotoEdit; A9, ©Arthur Gloor/Animals Animals; A10, A11, Brent Turner/BLT Productions; A17, ©Gerard Lacz/Animals Animals; A18, A19, ©KS Studios; A21 (tr), ©James H. Robinson/Animals Animals, (b), ©Larry Lipsky/Tom Stack & Associates; A22 (tl), ©Eric & David Hosking/Photo Researchers, Inc., (cl), ©Stephen J. Krasemann/DRK Photo, (br), ©Charlie Palek/Animals Animals; A23 (tr), ©Joe McDonald/Animals Animals, (bl), ©Stephen J. Krasemann/DRK Photo; A24 (tl), ©Andrew J. Martinez/Photo Researchers, Inc., (br), ©Stefan Meyers/Animals Animals; A25, ©Jim Tuten/Earth Scenes; A26, A27, KS Studios; A30-A31, ©Corbis-Bettmann; A33, ©Sylvain Grandadam/Photo Researchers, Inc.; A34 (tl), ©Darrell Gulin/DRK Photo, (bl), ©D. Cavagnaro/DRK Photo; A35, ©John Hyde/Bruce Coleman, Inc.; A36, © Juan M. Renjilo/Earth Scenes; A37, ©Renee Lynn/Photo Researchers, Inc.; A38, A39, KS Studios; A41 (tr), ©David Young-Wolff/PhotoEdit, (b), ©John Stern/Earth Scenes; A42, ©Holt Conter/DRK Photo; A43, ©David Young-Wolff/PhotoEdit; A44, ©Tim Davis/Photo Researchers, Inc.; A45, A46, A47, KS Studios; A50-A51, ©David Weintraub/Photo Researchers, Inc.; A53, ©L. Nielsen/Peter Arnold, Inc.; A54-A55, ©Michael Fairchild/Peter Arnold, Inc.; A56 (tl), ©James F. Housel/Tony Stone Images, Inc., (bl), ©Fred Whitehead/Animals Animals; A57 (tr), ©Bernard, G.I./Oxford Scientific Films/Animals Animals, (br), ©Richard Kolar/Animals Animals; A58, A59, KS Studios; A61 (tr), ©Tom Bean/DRK Photo, (b), ©T.A. Wiewandt/DRK Photo; A64, A65, ©Corbis/Bettmann; A67, KS Studios; A69, ©Robert Maier/Animals Animals; A70 (tl), ©Thomas Dressler/DRK Photo, (bl), ©Tom McHugh/Photo Researchers, Inc., (br), ©Corbis/Gary Braasch; A71(tr), ©Joe McDonald/Animals Animals, (bl),©John Gerlach/DRK Photo; A72 (l), ©Tom & Pat Leeson/DRK Photo, (r), ©Lynn M. Stone/DRK Photo, (b), ©Michael Fogden/DRK Photo, (blc), ©Maresa Pryor/Earth Scenes; A73, ©Douglas Peebles/Corbis; A75, A79, ©KS Studios; B2-B3, ©NASA/PhotoTake; B5, ©Matt Meadows; B10, B11, Doug Martin; B14 (tl, br), ©Photri, (tr), ©NASA/International Stock, (bl), NASA; B15 (tl, tr), NASA /International Stock, (bl), ©Photri, (br), NASA, (cr), ©Frank P. Rossotto; B18, B19, KS Studios; B26, B27, Matt Meadows; B30-B31, ©David L. Brown/Tom Stack & Associates; B36, ©Roland Seitre/Peter Arnold, Inc.; B37, ©Vince Streano/Corbis; B39, Platinum Studios; B41 (l), ©SuperStock, (cl), ©Alex Bartel/Science Photo Library/Photo Researchers, Inc., (cr), ©Francois Gohier/Photo Researchers, Inc., (tr), ©Grant Heilman/Grant Heilman Photography; B42, ©Grant Heilman/Grant Heilman Photography; B43, ©Corbis; B44 (tl), ©Kevin Schafer/Corbis, (bl), ©Kermani/Gamma Liaison Agency; B45 (l), ©Craig Aurness/Woodfin Camp & Associates, (cr), ©Roger Werths/Longview Daily News/Woodfin Camp & Associates; B46, B47, Ken Karp; B48, ©Tom Stack/Tom Stack & Associates; B49 (tl), ©Jim Steinberg/Photo Researchers, Inc., (tr), ©Ellen Dooley/Tony Stone Images, (bl), ©Luiz C. Marigo/Peter Arnold, Inc., (br), ©Kevin Schafer/Peter Arnold, Inc.; B50 (tl), ©John Kieffer/Peter Arnold, Inc., (b), ©Breck P. Kent/Earth Scenes; B51 (tl), ©Marty Stouffer/Earth Scenes, (br), ©Sylvain Grandadam/Photo Researchers, Inc.; B52, ©Comstock; B54, B55, KS Studios; B58-B59, ©Greg Vaughn/Tom Stack & Associates; B61, ©Steve Kaufman/Peter Arnold, Inc.; B62 (t), ©Lynn McLaren/Photo Researchers, Inc., (br), ©Barry L. Runk/Grant Heilman Photography; B63, ©Comstock; B65, ©David Young Wolff/Tony Stone Images, Inc.; B66, B67, Ken Karp; B69 (tr), Matt Meadows, (br), ©Runk/Schoenberger/Grant Heilman Photography; B72, ©Larry Lefever/Grant Heilman Photography; B73 (tl), ©Larry Lefever/Grant Heilman Photography, (cr), ©Grant Heilman/Grant Heilman Photography, (bl), ©Jim Strawser/Grant Heilman Photography; B75, Matt Meadows; B76, ©Jim Steinberg/Photo Researchers, Inc.; B77, ©Zig Leszczynski/Animals Animals; B78, ©Jim Richardson/Corbis; B79, ©Kevin Schafer/Peter Arnold, Inc.; B80, ©Thomas Kitchin/Tom Stack & Associates; B81, ©FPG International; B82, B83, Matt Meadows; B87, KS Studios; C2-C3, ©Stephen Johnson/Tony Stone Images; C5, ©Ed Pritchard/Tony Stone Images; C7, KS Studios; C8, ©Dana White/PhotoEdit; C9, ©G. Brad Lewis/Tony Stone Images; C10, C11, Studiohio; C14, Matt Meadows; C15, ©Ian O'Leary/Tony Stone Images; C16, First Image; C17, ©David Burnett/The Stock Market; C18, Studiohio; C21 (t), ©Frank Saragnese/FPG International, (b), ©Tom Tracy/The Stock Market; C24, C25, ©David Young-Wolff/PhotoEdit; C26, C27, Matt Meadows; C30-C31, ©Barbara Filet/Tony Stone Images; C33, ©Ken Biggs/The Stock Market; C34, ©Fergus O'Brien/FPG International; C37, Matt Meadows; C39, ©Schmid/Langsfeld/The Image Bank; C40, C41, Matt Meadows; C43 (tl), ©Vic Bider/Tony Stone Images, (br), ©Mike Hewitt/Tony Stone Images; C44, ©Matt Meadows; C45, ©David Young-Wolff/PhotoEdit; C46, KS Studios; C47, ©David Young-Wolff/PhotoEdit; C48, C49, KS Studios; C50-51, ©ZEFA Germany/The Stock Market; C55 (tl), ©Jon Gray/Tony Stone Images, (cr), Matt Meadows; C56, C57, Matt Meadows; C60-61, ©Aaron Haupt/PhotoResearchers, Inc.; C64 (cl), ©Marcia Griffen/Animals Animals, (br)Tim Courlas; C65, Studiohio; C68, C69, Ken Karp; C71 (tr), ©David Young-Wolff/PhotoEdit, (bl), ©Spencer Grant/PhotoEdit; C72 (cl), Matt Meadows, (br), KS Studios; C73, ©Tom Stewart/The Stock Market; C75, KS Studios; C76, C77, Matt Meadows; C80-81, ©Patrick Bennett/Corbis; C83, ©Corbis; C84, ©Calvin Larsen/Photo Researchers, Inc.; C85, Aaron Haupt; C86, C87, Matt Meadows; C88, Aaron Haupt; C89, ©Russell D. Curtis/Photo Researchers, Inc.; C91, ©Michael Newman/PhotoEdit; C93, ©Mason Morfit/FPG International; C96, KS Studios; C97, ©Fujifotos/The Image Works; C98, Matt Meadows; C101, ©Tony Freeman/PhotoEdit; C102, C103, Matt Meadows; C104, ©Tom Stack & Associates; C105, ©SuperStock; C106, C107, Studiohio; C111, KS Studios; D2-D3, ©Richard Hamilton Smith/Corbis; D6, ©Michael Newman/PhotoEdit; D7, ©David Young-Wolff/PhotoEdit; D8, ©John Neubauer/PhotoEdit; D10, Matt Meadows; D13, ©David Madison 1987; D14, ©Stephen McBrady/PhotoEdit; D15, ©Tony Freeman/PhotoEdit; D17, ©Myrleen Ferguson/PhotoEdit; D19, D26, D27, Matt Meadows; D30-D31, ©David Young-Wolff/PhotoEdit; D33, ©Myrleen Ferguson/PhotoEdit; D34, D35, D36, ©David Young-Wolff/PhotoEdit; D37 (cr), ©Alon Reininger/The Stock Market, (tr), ©Cindy Charles/PhotoEdit; D38, D39, D41, D42, Matt Meadows; D43, KS Studios; D44, D45, D46, D47, Matt Meadows; D49, KS Studios; D50, ©Tony Freeman/PhotoEdit; D51, ©Michael Newman/PhotoEdit; D52, ©Charles Thatcher/Tony Stone Images; D54, Matt Meadows; D55, ©Cassy M. Cohen/PhotoEdit; D58-D59, ©David Young-Wolff/PhotoEdit; D61, ©Mark C. Burnett/Photo Researchers, Inc.; D62, ©Ricardo Arias,

R39

Credits